Water Pollution and its Recent Challenges

Water Pollution and its Recent Challenges

Editor

Dr. Hashmat Ali

Environmental Science Research Unit
Post Graduate Department of Chemistry
S.K.M. University, Dumka – 814 101 (Jharkhand)

2012

DAYA PUBLISHING HOUSE®

New Delhi - 110 002

ISBN 9789351242307

Published by : **Daya Publishing House®**
A Division of
Astral International Pvt. Ltd.
– ISO 9001:2008 Certified Company –
4760-61/23, Ansari Road, Darya Ganj,
New Delhi - 110 002
Phone: 23245578, 23244987
Fax: (011) 23260116
e-mail : dayabooks@vsnl.com website : www.dayabooks.com

Laser Typesetting : **Classic Computer Services**
Delhi - 110 035

Printed at : **Chawla Offset Printers**
Delhi - 110 052

PRINTED IN INDIA

Preface

By far the most burning problem our country is facing today is the problem of water pollution. Selling out of bottled water like cakes and biscuits in running trains and buses as well as at every square, bus-stop, railway station and in the market; installation of water purifying devices like Aqua-guard in every affordable house and in every office, school, college and hospital as the most essential and unavoidable commodity throughout the length and breadth of our country is itself a testimony that water around us is polluted. Moreover, with the accelerating increase in population the problem is gaining momentum in its complexity in the same proportion day by day. It is needless to mention that all living creatures including plants whether terrestrial or aquatic depend on water for their survival. In the words of Prof. B.N. Panday – 'water is most precious and important natural resource given by the nature not only to humanity but to the entire living community of the globe as a whole.' While teaching about water the teacher may explain that water consists of two atoms of hydrogen and one atom of oxygen but he misses the vital relationship of earth and human beings in terms of water present in both. We are 95 per cent water before we are born, 70 per cent water when we are an adult. Therefore, we are mostly water. On top of that we are always surrounded by water. The loss of water content in the human body leads to his death. Similarly, depletion of water on the earth will lead to its death. Then the fate of man need not be mentioned.

All major rivers of our country are facing ominous problems of water pollution. Presence of 60,000 faecal cauliform bacteria per 100 ml in river Ganga has made its water 120 times more polluted than the water used in taking bath. Yamuna with 10,000 such bacteria per 100 ml is suffering from the same problem inspite of undergoing 15 years cleansing programme. River water of Jharkhand (Damoder, Panchet, Subarnrekha) are also contaminated with various pollutants mainly by iron, lead, cobalt etc.

Even undergroundwater is getting polluted and not safe to drink. This has created a high level of tension among the Ecoplanners. The quantity of fluoride and arsenic is far above the acceptable limit at several places all over the country including Jharkhand state. Dumka, Sahebganj, Dhanbad, Hazaribagh, Chatra, Garhwa, Simdega, Bokaro etc. are some of the districts of Jharkhand contaminated with fluoride and arsenic. The undergroundwater in most of the districts of Jharkhand has already been reported to contain high percentage of iron. Increasing number of deep bore wells especially in urban areas due to growing population has already started aggravating the problems. This has created two fold problems of keeping the water level much deeper coupled with high concentration of mineral contents in undergroundwater. Unscientific exploration of water and reducing number of groundwater recharge facilities due to land acquisition, land reclamation including reclamation of ponds and other wetlands have brought further complexity in the situation.

According to reports of Geological Survey of India, over exploitation of groundwater at Jharia is 106 per cent, at Godda it is 122 per cent, at Kanke and Jamshedpur it has reached upto 113 and 132 per cent respectively. 93 per cent and 95 per cent areas of Dhanbad and Ramgarh respectively have been declared as 'Danger Zone'. Similarly, 76 per cent areas of Chas have been declared as 'Semi-Danger Zone.'

So far as fall in underground table at different places of Jharkahnd is concerned, it may be presented in the Table 1.

Groundwater at many places in the State has been found to contain Fluoride beyond permissible limit (Standard level = 1 to 1.5 ppm). Similarly, in some areas high concentration of Iron in groundwater has been recorded. The standard acceptable limit in the case of iron is 2 ppm. Areas with contents of Fluoride and Iron are being presented in the Table 2.

Table 1

Place	Extent of fall	Place	Extent of fall
Chatra	47 feet	Simdega	10.6 meters
Pakur	48 feet	Lohardaga	11.7 meters
Palamau	48 feet	Dumka	11.8 meters
Giridih	49 feet	Bokaro	12.1 meters
Dhanbad	50 feet	Hazaribagh	12.3 meters
Chas	51 feet	Jamatara	12.5 meters
Ranchi	52 feet	Gumla	13.8 meters
Jamshedpur	56 feet	Singhbhum	13.8 meters
Godda	61 feet		

Table 2

Conc. of Fluoride (Standard Level = 1.0 to 1.5ppm)		Conc. of Iron (Standard Level = 2ppm)	
Place	Quantity	Place	Quantity
Pakur	1.21	Dhanbad	2.92
Bhandariya	1.59	Singhbhum	4.06
Chas (Bokaro)	2.14	Dumka	4.95
Pratappur (Chatra)	2.42	Simdega	19.77
Garhwa	7.66	Bokaro	21.31
		Garhwa	25.02

Problems of Arsenic are also to be found at different places of Jharkhand State including Santal Pargana region. Sahebganj, Rajmahal and Udhwa blocks of Sahebganj district have been found to be alarmingly contaminated with the presence of Arsenic to the level of 10 ppb or more. The presence of Arsenic has also been reported from districts like Dumka, Pakur, Godda etc.

River flowing through the coalfields of Jharkhand have been reported to carry Arsenic responsible for arsenic poisoning in downstream areas of West Bengal. The coalfields of Bachara and

Piprawar areas of Jharkhand have contaminated the water of Damodar and its tributary, the Safi. Arsenic contamination arises mainly due to the dumping of wastes from the coal mines along the river bed. Coals of these areas have been found to contain sufficient amount of arsenic.

As a result of fall in groundwater table and deep bore wells, the groundwater is contaminated with iron, fluoride and arsenic. The same water is used for irrigation by almost 70 percent of farmers. This results in the presence of these chemicals in high amount in the food grain produced by crops irrigated with such water. It is notable here that the groundwater pollution is difficult to detect and costlier as well as time-consuming in monitoring. So far as Jharkhand state is concerned, according to one estimation, the exploitation of groundwater mainly through deep bore wells has reached upto 30 times more since previous three years whereas about 90 per cent facilities of groundwater recharge have been stopped completely.

The availability of drinking water in the state has reduced to 1200 cubic meter from 5200 cubic meter. On an average the fall of groundwater level has been recorded to be about 52 feet. Under these conditions what to talk of pollution in water, an acute crisis of water is soon going to be emerged in the state if accurate and sincere measures are not taken up on priority basis.

In this regard, Prof. B.N. Pandey while presenting his paper entitled 'Is third world war inevitable for water crisis in 21st century' in the seminar perhaps rightly claims by saying- A new category of refugees is expected to emerge around 2025, the water migrant and future wars between or within nations will be fought on the issues of sharing of water. According to water professionals, due to almost exponential growth of human population there has been overexploitation of groundwater leading to decline in its level by 2-4 meters at many places and consequently the world thirst for water is likely to become one of the most pressing issues of the 21st century as global water consumption rose six folds between 1900 and 1995.

Water is affected mainly by Industrial wastes, Organic sewage wastes, Chemical fertilizers, Petroleum oil etc. The wastes discharged by various industries pollute the water recklessly. Apart from acids, alkalies, phenols, organic compounds and inorganic minerals, these industrial wastes contain compounds of highly toxic elements like Mercury, Lead, Arsenic, Cadmium etc. These industrial wastes if disposed of to lakes and rivers can kill the aquatic lives as well as make the water non-utilisable. While intake of Lead causes problems in breathing and nervous system and in the synthesis of haemoglobin, Mercury creates nervousness, fear, irritability, pessimism etc. Cadmium causes lung irritation, hypertension and anaemia. Wastewater from Chemical industry, Metal industry, Food processing industry, Textile industry, Petroleum industry, Leather industry, Nuclear Power Plants and Nuclear Research Laboratories causes damages to living bodies in so many ways. Wastewater affects human beings, animal lives and plants alike.

Effluents discharged from mine sites have seriously polluted the streams and undergroundwater of the concerned areas throughout the globe. Acid mine drainage, liquid effluents from coal handling plants, colliery workshops and mine sites and suspended solids from coal washeries have caused serious water pollution throughout the country and the world over thereby adversely affecting fish and aquatic life. If we talk of our state, the Damodar river, the major source of water in the region is perhaps the most polluted river in India. It receives wastes from many industries situated on its banks. A single coal washery has been found to discharge about 40 tonnes of fine coal into Damodar river everyday. There are as many as eleven coal washeries in the region with an annual installed capacity of 20.52 million tonnes. Today the Damodar river has turned into mere a sewage canal shrunken and filled with filth and rubbish, emanating obnoxious odour. Other major rivers of the region are also seriously polluted. The Karo river in West Singhbhum is polluted with red oxide from the iron ore mines of Noamundi, Gua and Chiria. The Subarnarekha river shows a different type of pollution even more hazardous than this. Metallic and dissolved toxic wastes from TISCO, Jamshedpur and HCL, Ghatshila and radioactive wastes from the Uranium mill and tailing ponds of the UCIL at Jadugora flow into Subarnarekha and its tributaries.

In the light of what has been described above it has become unavoidable for the Central and State Governments to frame stringent laws and take firm and adequate measures to protect water from its pollution as well as its crisis by formulating National Water Policy and State Water Policy under which meticulous care to safeguard the sanctity of water could be taken with strong vigil and sincere monitoring of all the aspects related to water pollution.

It is worth mentioning here the words of Dr Masaru Emoto, Chairman, International water for Life Foundation, Japan who elsewhere says- 'Water is almost living and as sensible as we are; good music can positively affect water; water can positively react to good or bad words, peace or war, love and prayer; water can copy what it was shown and as such we should love and respect water because words have vibrations and spirits'.

The utterance by Dr. Emoto is quite comprehensible and appreciable if we take the thing on spectroscopic level. If we utter good words to water we thereby emit favourable radiations (to water) which make alignment with the resultant (Internal) energy possessed by water molecules in the form of vibrational, rotational, electronic energies etc. and good comes out of it. Liking and disliking attitudes by water molecule for the incident energy were shown through the structures of water molecule photographed under different conditions by Prof. Pandey in the Seminar.

In the context of safeguarding the sanctity of water by formulating National and State Water Policies mentioned above, it is imperative to know that during past 25 years our country had time to time constituted so many Committees and Authorities to clean Ganga but none proved effective. Under 'Mission Clean Ganga' the Central Govt. again has pledged to make Ganga pollution-free by 2020 by completely banning over the flow of industrial and untreated wastes. Related to it a new Authority known as 'National Ganga River Basin Authority' has been constituted under the Chairmanship of our Prime Minister. Resolution has been taken by the Government under the River Basin Management Plan formulated by the joint efforts of seven IITs to develop the Ganga banks along Uttarakhand, U.P., Bihar and West Bengal to build Electric Funeral Houses, to develop Sewer network, to implant Sewer Treatment Machines etc. All these measures resolved by the Central Government are praiseworthy provided these are brought to manifestation from paper.

The significant ideas evolved out during the deliberations made by the scholars and the scientists in the prestigious National Seminar on 'Water Pollution and its Recent Challenges' held in the month of April 2011 include Control of eutrophication; Silt management; Management of toxic metals like Chromium, Mercury, Lead, Cadmium etc.; Control of Cyanides, Arsenic, Fluoride, Iron; Control of Pesticides and Insecticides; Water resource management by the removal of toxic elements, pollutants and germs; Rain water harvesting; Recharging and increasing the level of undergroundwater; Water treatment processes such as Softening, Deionisation, Dealkalisation, Reverse Osmosis etc.; Advanced oxidation process for the treatment of hazardous micro pollutants; Role of algae in water purification; Use of Bentonite mineral to purify water etc.

Discourses delivered in the Seminar finally confirm that presence of heavy metals in water is definitely dangerous. Unfortunately we are still unmindful of the long-term effects which these metals may finally tell upon our health. Long term exposure of even low levels of these heavy metals results in the development of many incurable disorders.

My special thanks and appreciation go to the scientists and exalted invited speakers whose deliverance in the two-day National Seminar (UGC Sponsored) opened a new vista for the students, scholars and teachers engaged in research activities in the field of water pollution. For, I heartily acknowledge the most unforgettable cooperation extended by University Grants Commission.

Prominent among the dignified speakers were Prof. D.C. Mukherjii, *Honorary Secretary, Indian Chemical Society;* Prof. B.N. Pandey, *President, Zoological Society of India;* Prof. Bharat Singh (Chemistry), *Allahabad University, Allahabad;* Prof. K.N. Singh (Chemistry), *BHU, Varanasi;* Prof. S. Sarkar (Chemistry), *IIT, Kanpur;* Prof. M.N. Majumdar (Chemistry), *University of Kalyani (W.B.);* Prof. Bihari Singh (Env. Sc.), *A.N. College, Patna;* Prof. I.N. Mishra (Botany), *L.N.M.U., Darbhanga;* Prof. H.C. Rai (Chemistry), *B.R.A. Bihar University, Muzaffapur;* Prof. Ashok Kr. Jha. (Chemistry), *Naugachhia, Bhagalpur;* Prof. M. Pathak (Chemistry), *Mizoram University,* Prof. B. Mitra (Geography), *University of Burdwan;* Prof. P.K. Verma (Zoology), *Godda College, Godda;* Prof. B.N. Mandal, *ISI, Kolkata;* Prof. R.N. Majumdar (Chemical Technology), *University of Kalyani* and many more.

Prof. D.C. Mukherjee, Prof. M.N. Majumdar and Prof. R.N. Majumdar, the trio from Kolkaka not only graced the occasion by their kind presence but also guided our team for the successful completion of the Seminar. I am highly obliged to the trio.

I have ventured to compile some of the most significant and updated research deliberations presented by the scientists and the noble scholars in the Seminar to benefit the Students, Research Scholars, Teachers, Industrialists, NGOs and the Scientists working in the field of water pollution and to aware generations to come.

Lastly I pay deep sense of gratitude to Prof. (Dr.) M. Basheer Ahmad Khan, the Hon'ble Vice Chancellor, S.K.M. University, Dumka (Jharkhand) for his constant inspirations that encouraged me to take up this academic challenge of holding the National Seminar in the capacity of Convener. The present volume is actually, in his honour, dedicated to him. I can never thank Dr. Khan enough.

I also pay my humble gratitude to Prof. (Dr.) P.K. Ghosh, our departmental Head for his guidance at each step in organizing the Seminar. I owe my special thanks to Prof. Ashok Kr. Jha of Naugachhia (Bhagalpur) who always had valuable suggestions for me when required. I also acknowledge the cooperations provided by my departmental colleagues, Dr. S.K. Singh, Dr. V.P. Sahay, Dr. N.K. Mandal, Dr. Rishikesh Kumar, Dr. Chandan Kumar, Mr. C. S. Azad and Dr. R.K. Bishen, Dr.S.P. Yadav, Mr. Mirtunjay Kumar, the Research scholars and also Post Graduate students.

I am also thankful to Mr. Anil Mittal, proprieter, Daya Publishing House, Delhi and Mr. Rajiv Kumar Mishra (Zoo.), A.N. Inter College, Dumka for taking keen interest in bringing out this book.

Finally, I pay my sincere thanks to God for His grace without which this book in the present form would not have come into existence. Praise be to Him, the Almighty.

Dumka (Jharkhand) **Dr. Hashmat Ali**

Contents

List of Contributors

A. Paul
Envirocheck, 189 and 190, Rastragura Avenue, Kolkata-700028

Anil Kumar
National Institute of Foundry and Forge Technology, Hatia, Ranchi.

Anil Kumar Verma
Department of Philosophy, S.K.M. University, Dumka (Jharkhand)

Ahmad Ali
Junior Research Fellow, Department of Philosophy, Patna University, Patna

Amardip Singh
Assistant Professor cum R. Scientist, Department of Research and Planning, XISS, Ranchi

Archana Pandey
Associate Professor, Department of Chemistry, C.M.P. Degree College, Allahabad-211002

Arshad Ali
Research Scholar, Department of Philosophy, L.N.Mithila University, Darbhanga, Bihar

Ashok Kumar Jha
Department of Chemistry, G.B. College, Naugachhia, Bhagalpur

Awantika Kumari
Research Scholar, P.G. Department of Biotechnology, T.M.B. University, Bhagalpur

B. Rakshit
Envirocheck,189 and 190, Rastraguru Avenue, Kolkata-700028

Bhartendu Prasad Rai
Assistant Professor, Department of Chemistry, P.R. College, Sonpur, Bihar

Bina Pathak
Department of Hindi, Deoghar College, Deoghar (Jharkhand)

Biswajit Mitra
Assistant Professor, Department of Geography, Guskara Mahavidyalaya, Guskara, Barddhaman (West Bengal)

C.S. Azad
Assistant Professor, Department of Chemistry, S.P. College, Dumka (Jharkhand)

Chanda Kumari
Cambridge Institute of Technology, Tatisilwai, Ranchi

Chandan Kumar
Assistant Professor, Department of Chemistry, S.P. College, Dumka (Jharkhand)

D.C. Mukherjee
Department of Chemistry and Environment, Heritage Institute of Technology, Anandapur, Kolkata-700107

Deep Narayan Singh
Assistant Professor, Department of Physics, S.K.M. University, Dumka (Jharkhand)

Gopal Pal
Department of Bengali, Deoghar College, Deoghar (Jharkhand)

G.P. Sinha
Assistant Professor, Sant Kabir College, Samastipur (Bihar)

H.N. Tiwari
Ex-Dean and Head Faculty of Law, University of Allahabad, Allahabad

Hashmat Ali
Post Graduate Department of Chemistry, S.K.M. University, Dumka (Jharkhand)

Jahangeer
Department of Applied Sciences and Humanities, National Institute of Foundry and Forge Technology, Hatia, Ranchi

K. S. Awasthy
Associate Professor and Head, Department of Zoology, K.K.M. College, Pakur (Jharkhand)

K. Swarnim
Department of Zoology, Ranchi University, Ranchi

L. N. Patralekh
Department of Botany, Deoghar College, Deoghar (Jharkhand)

M.K. Singh
National Institute of Foundry and Forge Technology, Hatia, Ranchi

M.N. Majumdar
Formerly Professor of Chemistry and Dean, Faculty of Science, University of Kalyani, Nadia (West Bengal)

M. Pathak
Department of Chemistry, Govt. College, Lunglei, Mizoram

Mahesh Kumar Singh
Department of Philosophy, Deoghar College, Deoghar (Jharkhand)

Mohammad Ali
Research cholar, Department of Philosophy, Patna University, Patna

N.K. Mandal
Assistant Professor, Department of Chemistry, S.K.M. University, Dumka (Jharkhand)

Neetu
Department of Applied Chemistry, CIT, Tatisilwai, Ranchi

Nilima Verma
Assistant Professor, A.S. College, Deoghar (Jharkhand)

Nitesh Raj
Assistant Professor, Department of Economics, S.P. College, Dumka (Jharkhand)

Nitin Kumar
Department of Applied Sciences and Humanities, NIFFT, Hatia, Ranchi

Nutan Kumari
Assistant Professor, S.M.R.C.K. College, Samastipur (Bihar)

P.K. Verma
Life Science Research Laboratory, Department of Zoology, Godda College, Godda (Jharkhand)

Paulomi Das
Envirocheck, 189 and 190, Rastraguru Avenue, Kolkata-700028

Poonam
Resource Person, Department of EWM, A.N. College, Patna

Poonam
Research Scholar, Faculty of Law, University of Allahabad, Allahabad

Prasanjit Mukherjee
Assistant Professor, Department of Botany, K.K.M. College, Pakur (Jharkhand)

R.K. Bishen
Assistant Professor, Department of Chemistry, J.J.S. College, Mihijam, Jamatara (Jharkhand)

R.K. Singh
Department of Applied Chemistry, CIT, Tatisilwai, Ranchi

Rishikesh Kumar
Assistant Professor, Department of Chemistry, S.P. College, Dumka (Jharkhand)

S.B. Chowdhury
Envirocheck, 189 and 190, Rastraguru Avenue, Kolkata-700028

Saiyad Rafat Imam
Research Scholar, P.G. Department of Chemistry, H.D. Jain College, Ara (Bihar)

Saurabh Dutta
Department of Botany, S.K.M. University, Dumka (Jharkhand)

Santosh Kumar Singh
Assistant Professor, Department of Chemistry, S.P. College, Dumka (Jharkhand)

Satya Prakash
Research Scholar, Cyto-Lab, Life Science, Department of Zoology, K.K.M. College, Pakur (Jharkhand)

Subrata Saha
Department of Applied Sciences and Humanities, National Institute of Foundry and Forge Technology, Hatia, Ranchi

Sushmita Boyra
Department of Botany, Deoghar College, Deoghar (Jharkhand)

Veena Kumari
Department of Zoology, Deoghar College, Deoghar (Jharkhand)

Chapter 1

Bioindicators of Water Pollution

Archana Pandey

ABSTRACT

The quality of water is supposed to be the main factor which controls the state of health and disease in both animals and man. Nowadays, the increasing use of the waste chemicals and agricultural drainage systems represent the most dangerous chemical pollution. The most important heavy metals from the point of view of water pollution are Zn, Cu, Pb, Cd, Ni and Cr. Some of these metals (*e.g.* Cu, Ni, Cr and Zn) are essential trace metals to living organisms, but become toxic at higher concentrations. Others, such as Pb and Cd have no known biological function but are toxic elements. There are urgent demands for comprehensive methodological approaches to evaluate the actual state of water pollution and to monitor their rate of change[1]. Plants, animals and microbes can serve as bioindicator to integrate their total water environment.

Introduction

To measure the presence and effect of pollutants in water, bioindicators have attracted a great deal of interest. The principle behind the biomarker approach is the analysis of an organism to their metal contents in order to monitor the metal excess in their tissues. Various aquatic organisms occur in river, lakes, ponds, seas and marines are potentially useful as biomarkers of metal pollutants.

Bioindicators or biological indicators are the organisms that are used to monitor the health of the environment. They can provide information about the quality of environment, can detect changes in the natural environment also. They can tell us about the cumulative effects of different pollutants like they indicate the presence of pollutants, their intensity of exposure as well as how long a problem may have been present.

According to Market *et al.*[2], "*a bioindicator is an organism (or part of an organism or a community of organisms) they contain information on the quality of the environment (or a part of environment).*" An ideal indicator must have the following characteristics[3] :

1 Taxonomic soundness (easy to be recognized by non specialist),
2. Wide or cosmopolitan distribution,
3. Low mobility (local indication),
4. Well known ecological characteristics,
5. Numerical abundance,
6. Suitability for laboratory experiments,
7. High sensitivity to environmental stress, or
8. High ability for quantification and standaridization.[4]

Change in environment may cause change in physiology, behaviour, population or chemical contents of the bioindicator.

Types of Bioindicators

If the water is polluted, following changes in animal bioindicators may be observed:

- ☆ There may be increase or decrease of population.
- ☆ There may be increase or decrease of size.
- ☆ There may be increase or decrease of number.
- ☆ There may be increase or decrease the concentrations of toxins in animals tissues.
- ☆ There may be deformities in animals.

Some of the important animal bioindicators of water pollution are as follows:

Fish as Bioindicator

Fish are the top of the aquatic food web and are consumed by humans which makes them important for assessing contamination[5]. Due to their relatively long life cycle and mobility, they can be good indicators of long terms (several years) effects and broad habitat conditions[6]. Fish communities respond significantly and predictably to almost all kinds of anthropogenic disturbances including eutrophication, acidification, chemical pollution, flow regulation, physical habitat alteration and human exploitation[7]. Their sensitivity to the health of surrounding aquatic environments form the basis for using fishes to monitor water degradation[8].

Dolphin (*Stenella coerulocalba*)

It has been observed that the dolphins of Mediterranean coast showed the high concentration of Hg in liver, kidney, skeltol muscles, stomach and intestine. Liver has the highest concentration and melon has the lowest[9].

Tilapia nilotica

This fish was used as metal bioindicator as it is highly sensitivity to toxic effect[10]. Fist samples were collected from Nasser Lake (Egypt). The fish tissues includes muscle, gills, stomach, intestine, liver and scales. The fish ages were 1, 1.5, 2, 2,5 and 3 years. This study results that fish scales exhibited the highest concentrations of Cd, Pb, Co, Cr, Ni and Sr (0.088, 0.95, 0.29, 0.30, 0.25 3.21 microgram/gram respectively). The whole fish contains the higher concentrations of the studied metals compared to the previous study by Awadallah *et al.*[11] in the same fish from Nasser Lake and this means the increase in metal pollution in Lake water as the results of man activities.

Japanese scientists[12] used the fish *Tribolodon hakonsis* as bioindicator for arsenic accumulation in Usoriko lake. It was observed that large amount of arsenic was accumulated in the eye tissues. It means lake water contains relatively large amount of arsenic.

Cat Fish (*Clarias lazerd*)

It is considered as promising bioindicator for assessing water intoxication. The impact of metal pollution in the Nile and Delta lakes were studied in Cat fish[13]. Metals were accumulated most by the liver, less by muscles and least by serum. Iron showed the highest concentration in liver, muscles or serum compared to other metals. Even after the refinement of surrounding water, fish seem to uphold the metals that have been previously ingested. Mercury seems to eventually reside in muscles rather than in liver.

Seals (*Phoco vitulina*)

Some seals were found dead in Netherlands. High level of mercury has been detected in liver and brain[14]. As most of the mercury in marine fish is present as methyl mercury, the high levels in seal brain suggested that these animals are affected by the toxic action of methyl mercury compounds.

Turtles

Along the Atlantic coasts of France, cadmium, copper and zinc have been found in some tissues and organs of Logger head and Leather back turtles[15]. The pancreas of Leather back turtles exhibited the highest metal concentration which is very surprising for an organ which does not play a role in the detoxification process. The Cadmium concentration is highest in Leather back turtles. The main source of Cadmium for marine turtles is probably the food. The leatherback turtles are known to feed mainly on jelly fish in this area. Ten times higher cadmium concentration have been determined in jelly fish compared to fish and Leatherback turtles eat great quantity of jelly fish.

Crabs (*Carcinus maenas* L.)

The crab was exposed to a range of dissolved concentrations of Zn, Cu and Cd for 21 days in artificial sea water. Cd and Zn appear to be accumulated by crab at all exposures[16].

Insects

Aquatic insects can be used as bioindicator of water pollution. The family Chironomidae (Insect : Diptera) is the most abundant group among aquatic insects. Due to its short life cycle, small size and high density, different species of Chironomides have been used in acute toxicity tests. Behavioral responses to various heavy metal agents could be used as an indicator of environmental degradation. Taelwon kim and Tae - Soo - Chon[17] used the larvae of *Chironomus flaviplumes* to explore the possibility of using the behavioral responses as a bioindicator of heavy metal pollution of water. Under high concentration of copper, larvae increased outshelter behavior.

Frogs

Frogs and other amphibians live most of their lives in water making them very sensitive to water born chemicals which are absorbed through amphibians permeable skin. For this reason frogs often give the first warning of toxic pollutants by DDT and gammaxene. Along with these pesticides, synthetic hormones or hormone like substances, such as the plastic in baby bottles and the pesticides in crop sprays, contain chemicals that mimic the action of hormones. The effect of these hormones are most critical in early development *e.g.* some tadpoles look like big and healthy but they never develop into frogs. These substances inhibit their thyroid gland so they are stuck into the tadpole stage[18].

Marine Birds (Sea birds)

Sea birds are useful bioindicators of coastal and marine pollution[19]. Sea birds are defined as birds that spend a significant proportion of their life in coastal or marine environments. *Advantages* of using sea birds as bioindicators are as follows[20]:

1. *Top-predators* - high on food chain resulting in biomagnification.
2. *Long-lived*- time for bioaccumulation
3. *Philopatric* - allows for resampling of inviduals from year to year.
4. *Often Colonial* - large sample size concentrated in one place.
5. *Conspicuous* - easy to find for sampling
6. Large in size - do not require compositing
7. Wide spread distribution.
8. Many species are abundant.
9. Integrate over time and space.

There are some *disadvantages* also *e.g.*:

1. Gather food over large area.
2. Most species are migratory
3. Some species are threatened or endangered.

The major groups of pollutants are pesticides, chorinated hydrocarbons, metals and petroleum products. Since many of these contaminants are stored in tissues, the tissue level can be used as biomarker. Tissues usually collect in seabirds include blood, liver, kidney, brain and muscle.

It is not always possible to collect tissues due to the threatened or endangered status of species. In such cases, feathers are used as indicators. It has been demonstrated[21] that there is a high correlation between levels of mercury in the diet of seabirds and levels of mercury in their feathers, thus seabird feathers can be used as indicators of water pollution.

Lead poisoning showed drooped wings, loss of apetite, lethargy, weakness, tremors, green feces, impaired locomotion and lowered reproductive success. The common sea birds used as bioindicators are-*young common terns* (*Sterna hirundo*) and *herring gulls* (*Larus argentatus*). Lead, mercury, cadmium are found in feathers. Some species generally have higher levels of mercury (*e.g.* skimmers)[22].

Cory's Shearwater (*Colonectris diomedea*) is a long lived sea bird found in warm marine water from temperate to sub-tropical zones of the North Atlantic and Mediterranean. This seabird is used as bioindicator of sea water pollution with heavy metals from different coasts. Another seabird - *Red - billed Gulls* (*Larus novaehollandia,* scopulin) was used for monitoring of Hg level in the sea from New Zealand[23].

Shrimps and Oysters

These are also used as bioindicator of heavy metal pollution in sea water. White shrimps are very susceptible to the toxic effect of Zn[24]. Similarly Oysters from Elcho Island (Australia) showed high deposition of Fe, Cd, Zn, Mn, Cu, Ni and Pb[25].

Benthic Macroinvertebrates[26]

Aquatic invertebrates live in the bottom parts of water. They are also called benthic macroinvertebrates or benthos (benthic = bottom, macro - large, invertebrate = animal without a backbone) and are good indicators of watershed health because they:

1. Live in the water for all or most of their life.
2. Are easy to collect.
3.. Are easy to identify in a laboratory.
4. Often live for more than one year
5. Have limited mobility

Some benthos are found more often in larger amounts in clean water or unpolluted organic wastes. Without too much organic matter, the water usually have lots of oxygen for the benthos *e.g.* stoneflies are often considered as clean water benthos but worms and midges are the indicator of dirty water of river and streams.

Hypogean Crustaceans

Hyporheic is the zone located at the interface between surface and true groundwater. At this level there is an exchange of water, nutrients, organic matter and organisms between surface and groundwater, also there is a transfer of pollutants from surface river to groundwater table. Groundwater crustaceans and specially its stygobite fraction (species developing their entire life cycle exclusively in groundwater) are good predictors for the water quality. They are highly sensitive to any disturbances in their environment.

Stygobites displays specific adaptation *e.g.*

- ☆ *Morphologic Character* (*i.e.* blind, non pigmuted small and elongated body shape)
- ☆ *Metablic* (*i.e.* low metabolic rate, long life low reproduction rate)
- ☆ *Ecological* (narrower tolerance range)

Because of this strict specialization they are highly sensitive to any disturbances in their environment (both quality and quantity) and consequently their risk to be threatned are higher that turn to greater chances of extinction.

Some of the Hypogean crustaceans are: Cyclopoida, harpacticoida, acarina etc.

References

1. Rosenberg, D.M. and Resh, V.H. (Eds) (1993). *Freshwater Biomonitoring and Benthic Macroinvertebrates*. Chapman and Hall. New York.
2. Market, B., Wappelhorst, O., Weckert, V., Herpin, U., Siewers, U. and Friese, K. (1999). The use of bioindicators for monitoring the heavy-metal status of the environment. *Journal of Radioanalytical Nuclear Chemistry*, 240(2): 425–429.

3. Market, B., Breure, T. and Zechmeister, H. (Eds) (2003). *Bioindicators and Biomonitors: Principles, Concepts and Applications.* Elsevier, Amsterdam.

4. Hilty, J. and Merenlender, A. (2000). Faunal indicator taxa selection for monitoring ecosystem health. *Biological Conservation,* 92: 185–197.

5. Harris, J.H. (1995). The use of fish in ecological assessments. *Aust. J. Ecol.,* 20: 65–80.

6. Karr, J.R. (1981). Assessment of biotic integrity using fish communities. *Fisheries,* 6(6): 21–27.

7. Ormerod, S.J. (2003). Current issues with fish and fisheries: editor's overview and introduction. *Journal of Applied Ecology,* 40: 204–213.

8. Fausch, K.D., Lyons, J., Karr, J.R. and Anyermeir, P.L. (1990). Fish communities as indicators of environment degradation. *American Fisheries Society Symposium,* 8: 123–144.

9. Andre, J., Boudou, A., Ribeyre, F. and Bernhard, M. (1991). comparative study of mercury accumulation in dolphins (*Stenella coeruleoalba*). *Sci. Total Environ.,* 104: 191–209.

10. Patin, S.A. (1984) Tilapia as a bio-assay organism in toxicological studies. Biogeochemical and toxicological studies of water pollution. *Moskva (USSAR) Vniro,* pp. 39–46.

11. Awdallah, R.M., Mohamed, A.E. and Gaber, S.A. (1985). Determination of trace elements in fish by instrumental neutron activation analysis. *J. Radional. Nucl. Chem. Lett.,* 95(3): 145–154.

12. Takatsu, A. and Uchiumi, A. (1998). Abnormal arsenic accumulation by fish living in a naturally acidified lake. *Analyst,* 123: 73–75.

13. Adham, K.G., Hassan, J.F., Taha, N. and Amin T.H. (1999). Impact of hazardous exposure to metals in the Nile and Delta lakes on the Cat fish, *Clarias lazera. Environmental Monitoring and Assessment,* 54: 107–124.

14. Koeman, J.H., Peeters, W.H.M., Koudstaal-Hol, C.H.M., Tjioe, P.S. and De Goeiji, J.M. (1993). Mercury-selenium correlations in marine mammals. *Nature,* 245: 385–386.

15. Caurant, F., Bustamante, P., Bordes, M. and Miramand, P. (1999). Bioaccumulation of cadium, copper and zinc in some tissues of marine turtles along the French Atlantic coasts. *Marine Pollution Bulletin,* 38(12): 1085–1091.

16. Rainbow, P.S. (1985). Accumulation of Zn, Cu and Cd by Crabs. *Science,* 21(5): 669–686.

17. Taewon, K and Chon. Tae-Soo, Chirononids as bioindicator of water quality, Taewon Kim's homepage.

18. Hayes, T. (1999). *The Exploratorium: Frogs: Inside the Lab and Out in the Field.*

19. Furness, R.W. and Camphuysen, K.C.J. (1997). Sea birds as monitors of the marine environment. *ICES Journals of Marine Science,* 54: 726–723.

20. Burger, J. and Gochfeld, M. (2004). Marine birds as sentinels of environmental pollution. *Eco Health,* 1: 263–274.

21. Monteiro, L.R. and Furness, R.W. (1995). Sea birds as monitor of mercury in the marine environment. *Water, Air and Soil Pollution,* 80: 831–870.

22. Burger, J. and Gochfeld, M. (2000). Metal levels in feathers of 12 species of seabirds from Midway Atoll in the northern Pacific Ocean. *The Science of the Total Environment,* 257: 37–52.

23. Furness, R.W. and Lewis, S.A. (1990). Mercury levels in the plumage of Red-billed Gulls of known sex and age. *Environm. Pollt.,* 63: 33–39.

24. Magliette, R.J., Doherty, F.G., Mackinney, D. and Venkatarmani, L. (1995). Need for environmental quality guidelines based on ambient fresh water qualikty. *Bull Environ. Contam. Toxicol.,* 54: 532–626.

25. Szefer, P., Jkuta, K. and Geldon, J. (1997). Distribution of trace metals in pacific oyster. *Bull. Environ. Contam. Toxicol.,* 58: 108–114.

26. Invertebrates as Bioindicators – Biological Indicators of watershed health. US E PA.

27. Aguado, J. (2011). Hypogean crustaceans as bio-indicators for groundwater pollution. Publicado por.

Chapter 2

A Global Problem of Arsenic in Drinking Water and its Mitigation

M.K. Singh and Anil Kumar

ABSTRACT

Heavy metals contamination in drinking water is a serious problem throughout the world. Especially, arsenic is a global problem in drinking water affecting countries of all five continents. The most serious damage to health has been reported from Bangladesh and West Bengal, India. UNICEF estimated that 12 million people in Bangladesh were drinking arsenic contaminated water in 2006, and the number of people showing symptoms of arsenicosis was 40,000, but could rise to one million [1]. Arsenic contaminated groundwater is used in many countries on all continents as drinking water. Hundreds of millions of people, mostly in developing countries, daily use drinking water with arsenic concentrations several times higher than the World Health Organization recommended limit of 10mg/L. Consequently, ingestion, inhalation or skin adsorption, gastrointestinal symptoms, disturbance of cardiovascular and nervous systems functions occurs. Arsenic contamination is largely a natural phenomenon and their preventive measures may be public awareness campaigns, Mass media for publicizing the problem, identifying the arsenic contaminated sources, sharing arsenic free point sources and removal of arsenic at house hold level and treatment plant installed at community level. This Chapter highlights the effect of arsenicosis on human health and its ways of mitigation.

Keywords: *Arsenic, Drinking water, Human health and Mitigation.*

Introduction

The acute toxicity of arsenic at high concentrations has been known about for centuries. It has a strong adverse effect on health, even very low arsenic concentrations with long-term exposure. The presence of arsenic in drinking water is difficult to detect without complex analytical techniques because no any change in taste, odour or visible appearance of water. Hundreds of millions of people, mostly in developing countries, daily use drinking water with several times higher arsenic concentrations than the reported limit 10 µg/L of water by World Health Organization (WHO). The effects of arsenicosis are serious and lead to several forms of cancer and became a global problem due to affecting countries in all five continents. The most serious damage to health has taken place in Bangladesh and West Bengal, India. In 2000, a WHO report [2] described the situation in Bangladesh as: "the largest mass poisoning of a population in history beyond the accidents at Bhopal, India, in 1984, and Chernobyl, Ukraine, in 1986." In 2006, UNICEF reported that 4.7 million (55 per cent) of the 8.6 million wells in Bangladesh [1] had been tested for arsenic of which 1.4 million (30 per cent of those tested) had been painted red, showing them to be unsafe for drinking water due to presence of 50ppb [1]. UNICEF estimates that 12 million people in Bangladesh were drinking arsenic Contaminated water in 2006, and the number of people showing symptoms of arsenicosis Were 40,000, but could rise to one million [1]. The only ways to counteract the effects of arsenic contaminated water are to switch to unpolluted sources or to remove the arsenic before water is consumed. Use of alternative deep ground or surface water sources is expensive and not a solution in the short term for the most affected populations in rural areas. Rainwater harvesting has high investment costs, brings its own potential water quality problems and is of doubtful suitability in Countries, such as Bangladesh, where rainfall is seasonal. Sustainable production of arsenic free water from a raw water source that contains arsenic is very difficult due to the limited efficiency of conventional water treatment technologies, the high cost and complexity of advanced treatment and the generation of large volumes of waste streams that contains arsenic. The situation is most difficult in rural areas in developing countries, where arsenic contaminated groundwater is the only drinking water source. In such areas, where centralized systems usually do not exist, arsenic removal technologies suitable for Centralized water supply systems are not applicable. Efforts are being made to develop effective household treatment systems, but these too have proved problematic, both technically and operationally. This Chapter provides an up-to-date overview covering the extent of the problem of arsenic in drinking water, related health and social problems, arsenic chemistry, analysis and standards, arsenic removal processes and systems, and social and institutional issues associated with mitigation of the problem.

Health and Social Problems with Arsenic in Drinking Water

Inorganic arsenic is more toxic than organic arsenic, which affects gastrointestinal symptoms, disturbance of cardiovascular and nervous systems functions (*e.g.* muscle cramps, heart complains) or even death. Inorganic arsenic is present as trivalent in compounds. The numbers of acute toxic arsenic compounds are given in Table 2.1 [3].

Table 2.1: Acute Toxicity for Different Arsenic Compounds

Arsenic Form	*Oral LD50 (mg/kg body weight)*
Sodium Arsenite	15–40
Arsenic Trioxide	34
Calcium arsenate	20–800
Arsenobetane	>10,000

However, a very low concentration of arsenic in drinking water is also a health hazard for a long duration exposure [1, 4-7]. The first visible symptoms caused by exposure to low arsenic concentrations in drinking water are abnormal black-brown skin pigmentation known as *Melanesia* and hardening of palms and soles known as *keratosis*. If the arsenic intake continues, skin de -pigmentation develops resulting in white spots that looks like raindrops (medically described as *leukomelanosis*). In a clinical study conducted in West Bengal on a population exposed to high levels of arsenic in drinking water, 94 per cent had such "raindrop" pigmentation [8]. Palms and soles further thicken and painful cracks emerge. These symptoms are described as *hyperkeratosis* and can lead on to skin cancer [7]. Other cancers are also caused by long-term exposure to arsenic in drinking water. Arsenic may attack internal organs without causing any visible external symptoms and elevated concentrations in hair, nails, urine and blood can be an indicator of human exposure to arsenic before visible external symptoms [9]. The disease symptoms caused by chronic arsenic ingestion are called *arsenicosis* and developed only after more than ten years of exposure to arsenic contaminated water. Symptoms may develop to take 20 years of exposure for some cancers. Long-term ingestion of arsenic in water can first lead to problems with kidney and liver function, and then to damage to the internal organs including lungs, kidney, liver and bladder. Arsenic can disrupt the peripheral vascular system leading to gangrene in the legs, known in some areas as black foot disease. First reported symptoms of chronic arsenic poisoning were observed in China (province of Taiwan) in the first half of twentieth century. The International Agency for Research on Cancer has concluded that: "There is sufficient evidence in humans that arsenic in drinking-water causes cancers of the urinary bladder, lung and skin" [10]. The U.S. Environmental Protection Agency has estimated that the lifetime risk of skin cancer for individuals who consumed 2 liters of water per day at 50 µg/L of Arsenic [11]. UNICEF reported 40,000 confirmed cases of arsenicosis in Bangladesh [1]. Smith *et al.* in 2000 reported that the concentration of arsenic 500 mg/l or more caused might die from arsenic related cancers [2]. The presence of iron and manganese in water can reduce exposure to arsenic through adsorption and precipitation into iron and manganese precipitates and also delay the development of symptoms through lowering intake drinking water and consuming food rich in proteins and vitamins. There is no medical treatment for this disease and the only prevention is to stop ingesting arsenic.

Guidelines and Standards

Because of the proven and widespread negative health effects on humans, in 1993, the WHO lowered the health-based provisional guideline for a "safe" limit for arsenic concentration in drinking water from 50 µg/L to 10 µg/L [12,13]. The guideline value for arsenic is provisional because there is clear evidence of hazard but uncertainty about the actual risk from long-term exposure to very low arsenic concentrations [12,13]. The value of 10 µg/L was set as realistic limit taking into account practical problems associated with arsenic removal to lower levels. The WHO provisional guideline of

10 µg/L has been adopted as a national standard by most countries, including Japan, Jordan, Laos, Mongolia, Namibia, Syria and the USA, and by the European Union (EU). Implementation of the new WHO guideline value of 10 µg/L is not currently feasible for a number of countries strongly affected by the arsenic problem, including Bangladesh and India, which retain the 50µg/L limit. Other countries have not updated their drinking water standards recently and retain the older WHO guideline of 50 µg/L [6]. These include Bahrain, Bolivia, China, Egypt, Indonesia, Oman, Philippines, Saudi Arabia, Sri Lanka, Vietnam and Zimbabwe. The most stringent standard currently set for acceptable arsenic concentration in drinking water is by Australia, which has a national standard of 7 µg/L.

Worldwide Extent of Arsenic Problem

Inorganic arsenic found in groundwater is in most cases of geological origin. It is highly soluble and mobile in water [13]. Typical arsenic concentrations in groundwater are very low and in most cases below 10 µg/L. Elevated arsenic concentrations up to 5,000 µg/L are typically found in areas with active volcanism, geothermal waters, sedimentary rocks and in soils with a high concentration of sulphides (*e.g.* arsenopyrite). Reported arsenic concentrations above accepted standards for drinking water have been demonstrated in many countries in all continents is given in Table 2.2.

Table 2.2: Arsenic Affected Groundwater Countries

Asia	Bangladesh, Cambodia, China (including provinces of Taiwan and InnerMongolia), India, Iran, Japan, Myanmar, Nepal, Pakistan, Thailand,Vietnam
Americas	Alaska, Argentina, Chile, Dominica, El Salvador, Honduras, Mexico, Nicaragua, Peru, United States of America
Europe	Austria, Croatia, Finland, France, Germany, Greece, Hungary, Italy,Romania, Russia, Serbia, United Kingdom
Africa Pacific	Ghana, South Africa, Zimbabwe, Australia, New Zealand

The large scale of the arsenic problem is mainly in the alluvial and deltaic aquifer of Bangladesh and West Bengal, where millions of people drink water with high levels arsenic. British Geological Survey (BGS 1999) reported that 46 per cent of shallow wells (up to 150 meters), arsenic concentrations exceed the WHO guideline of 10 μg/L and Up to 57 million people were daily exposed to arsenic levels in drinking water that exceeded 10 μg/L, in some cases as high as 2,500 μg/L [14]. UNICEF reported in 2006 that 1.6 million (32 per cent) of the 5 million tube wells so far tested were found to contain arsenic above 50 μg/L [1]. An additional six million people in West Bengal (India) are believed to be exposed to arsenic levels of between 50 and 3,200 μg/L [13,15]. Most of the affected population in Bangladesh and West Bengal live in rural areas characterized by an absence of centralized water supply systems. Many millions of people in this region are drinking arsenic affected water daily, thousands have already been identified with arsenic related symptoms, and the fear is that their numbers could grow exponentially. In Europe, the arsenic problem is most alarming in Hungary, Serbia and Croatia. A report on groundwater quality in Hungary that the drinking water contain arsenic concentrations several times higher the WHO and EC guidelines [16]. Recent legislation directs water supply companies in Hungary to meet the EC drinking water directives, including ensuring that arsenic concentration are below 10 μg/L, by 2009. Fulfilling this requirement will be a major challenge for the water supply companies in this country. The full extent of the problem in Serbia is not yet known. Mexico, United States, Chile and Argentina are most affected by the arsenic problem in the Americas. It has been estimated that at least four million people are exposed to arsenic level > 50 μg/L in Latin America alone [17]. Extremely high arsenic concentrations in order of milligrams per litre were found in some wells in Latin America, including Bolivia and Peru. Levels as high as 5,000 μg/L have been recorded in Argentina [17] and reaching as high as 11,500 μg/L in some wells in Cordoba Province [15]. The pattern of arsenic presence in different wells, especially in the sedimentary aquifer with elevated arsenic concentrations (*e.g.* Bangladesh and Hungary) can be very irregular. Two nearby wells with similar depths can show a large variation in arsenic concentrations presumably due to a difference in sedimentary characteristics. It has also been found that arsenic concentration in a well can strongly increase within a few years of groundwater abstraction. At the same time, standards for an acceptable arsenic level in drinking water have become more stringent. It is therefore expected that arsenic problem in drinking water will be increasing in coming years and new countries will be identified.

Figure 2.1: Countries where Arsenic has been Reported in Ground or Surface Waters

Pentavalent arsenate (As (V)) in drinking water. Organic arsenic species are found in various forms as monomethyl arsenic acid (MMAA), dimethyl arsenic acid (DMAA), and arseno-sugars, which are very much less harmful to health, and are readily eliminated by the body. Arsenic is perhaps unique among the heavy metalloids and oxy-anion-forming elements (*e.g.* arsenic, selenium, antimony, molybdenum, vanadium, chromium, uranium, rhenium) in its sensitivity to mobilisation at the pH values (6.5 – 8.5) found in groundwater and under both oxidizing and reducing conditions. The valency and species of inorganic arsenic are dependent on the redox conditions and the pH value of the water. In general, arsenite, the reduced trivalent form [As (III)], is normally found in groundwater (assuming anaerobic conditions) and arsenate, the oxidized pentavalent form [As (V)], is found in surface water (assuming aerobic conditions), although the rule does not always hold true for groundwater. Some groundwaters have been found to have only As (III), others only As (V), while in some others both forms have been found in the same water source [19-22]. As (V) exists in four forms in aqueous solution based on pH namely H_3AsO_4, $H_2AsO_4^-$, $HAsO_4^{2-}$ and AsO_4^{3-}. Similarly, As (III) exists in five forms: $H_4AsO_3^+$, H_3AsO_3, $H_2AsO_3^-$, $HAsO_32^-$, and $AsO_3 3^-$. The ionic forms of As (V) dominate at pH >3, and As (III) is neutral at pH <9 and ionic at pH >9. Conventional treatment technologies used for arsenic removal, such as iron removal by aeration followed by rapid sand filtration rely on adsorption and co-precipitation of arsenic to metal hydroxides. Therefore, the valency and species of soluble arsenic have significant effect on arsenic removal [23]. The toxicity and mobility of arsenic varies with its valency state and chemical form. As (III) is generally more toxic to humans and four to ten times more soluble in water than As (V) [24-25]. Chemical speciation is a critical element for the removal of arsenic. Negative surface charges facilitate removal by adsorption, anion exchange, and co-precipitation processes. Since the net charge of As (III) is neutral at natural pH levels (6-9). This form is not easily removed. However, the net molecular charge of As (V) is negative (-1 or -2) at natural pH levels, enabling it to be removed with greater efficiency. Conversion of As (III) to As (V) is a critical element of any arsenic treatment process. This conversion can be accomplished by adding an oxidizing agent such as chlorine or permanganate.

Analysis of Arsenic

Determination of the speciation and concentration of arsenic in water is the first step in the assessment of the extent and severity of arsenic contamination in any given area. Arsenic in water can be measured in the laboratory or in the field using one of several field test kits.

Field Analysis (Test Kits)

Field test kits have been used extensively to test for arsenic in groundwater. Some commonly used arsenic measurement test kits are shown in Table 2.3.

Table 2.3: Commonly Used Arsenic Test Kits

Sl.No.	Test Kit	Range of Measurement (μg/L)
1.	MERCK (Germany)	5 – 500
2.	HACH (USA)	10 – 500
3.	Quick (USA)	10 – 1000
4.	AIIH and pH Kit (India)	Yes/No
5.	NIPSOM (Bangladesh)	10 – 700
6.	GPL (Bangladesh)	10 – 2500
7.	Arsenator (UK)	<10 – 500

The current baseline methodology involves a variety of technologies, all variations on the "Gutzeit" method. These involve treating the water sample with a reducing agent (*e.g.* zinc) that separates the arsenic by transforming arsenic compounds in the water into arsenic trihydride (arsine gas AsH_3). Arsenic trihydride diffuses out of the sample where it is exposed to a paper impregnated with mercuric bromide. Reaction with the paper produces a highly coloured compound. By comparing the colour of the test strip to a colour scale provided with the kit, the amount of arsenic in a sample can be estimated [26]. A photometer test kits with an electronic display to measure accurately the colour on the paper. Field-test kits can detect the presence of arsenic at high concentrations. However, test kits are generally inaccurate for detecting lower concentrations that are still of concern for human health [7]. Many of the test kits claim accuracy down to 0.01 mg/L.

Laboratory Analysis

Several accurate measurement methods of concentration of arsenic are available, which is relevant to the health, requires laboratory analysis, using sophisticated and expensive techniques [7]. The most common of these methods include:

1. Atomic absorption spectroscopic method: (a) Hydride generation atomic absorption (AAS – HG) and (b) Electro thermal atomic absorption (AAS – GF)
2. Silver diethyldithiocarbamate method (SDDC)
3. Inductively coupled plasma (ICP) method: (a) Mass spectrometry (ICP-MS) and (b) Atomic emission spectrometry (ICP-AES)
4. Anodic stripping voltammetry (ASV)

The hydride generation-atomic absorption method, which converts arsenic compound to their hydrides, although the electro-thermal method is simpler in the absence of interferences. The silver diethyl dithiocarbamate method, in which arsine is generated by sodium borohydride in acidic solution, is applicable to the determination of total inorganic arsenic when interferences are absent and the sample contains no methyl arsenic compounds. The diethyl dithiocarbamate method is also used for identifying and quantifying two arsenic species (arsenite and arsenate). The ICP method is useful at higher concentrations (greater than 50 µg/L). ASV is useful to quantify free, dissolved arsenic in aqueous samples. The chief advantage is that this technique does not require expensive instrumentation and is therefore useful for field analysis [27-28]. Analytical methods currently approved by US Environmental Protection Agency (USEPA) for analysis of arsenic in drinking water and detection limits are summarized in Table 2.4.

Arsenic Removal Technologies

Several treatment technologies have been adopted to remove arsenic from drinking water under the laboratory and field conditions. The major mode of removing arsenic from drinking water is a physical-chemical treatment. These are the following technologies for removing arsenic from drinking water:

- Precipitation processes, including coagulation/filtration, direct filtration, coagulation assisted microfiltration,

Table 2.4: Approved Analytical Methods by USEPA for Analysis of Arsenic in Drinking Water

Method		*Technique*	*Lowest Limit of Detection (µg/L)*
Multi-analyte Methods	EPA 200.8	ICP-MS	1.4
	EPA 200.7	ICP-AES	8
	SM 3120 B	ICP- AES	50
Single-analyte Methods	EPA 200.9	AAS-GF	0.5
	SM 3113 B	AAS-GF	1
	ASTM D 2972-93 Test method C	AAS-GF	5
	SM 3114 B	AAS-HG	0.5
	ASTM D 2972-93 Test method C	AAS - HG	1

Adapted from USEPA (1999).

enhanced coagulation, lime softening, and enhanced lime softening.

- ☆ Adsorptive processes, including adsorption onto activated alumina, activated carbon and iron/manganese oxide based or coated filter media
- ☆ Ion exchange processes, specifically anion exchange
- ☆ Membrane filtration, including nanofiltration, reverse osmosis and electrodialysis reversal
- ☆ Alternative treatment processes, especially greensand filtration
- ☆ *In situ* (sub-surface) arsenic removal [29-30]
- ☆ Biological arsenic removal [31]

Traditional treatment processes technologies like coagulation/ filtration, lime softening, iron/manganese oxidation, and membrane filtration have been tailored to improve removal of arsenic from drinking water in water treatment plants. Technologies such as ion exchange, manganese greensand filtration and adsorption on activated alumina have been employed in small and domestic systems. Innovative technologies, such as permeable reactive barriers, biological treatment, phytoremediation (using plants), and electrokinetic treatment, are also used to treat arsenic-contaminated water. However, many of these techniques are at the experimental stage and some have not been demonstrated at full-scale. Also, although some of these processes may be technically feasible, their cost may be prohibitive [32-33]. For these reasons, only the most common methods like precipitation processes, adsorption processes, ion exchange and membrane filtration are used for arsenic removal.

Arsenic Removal Systems

The arsenic removal technologies outlined above can be employed either in centralized treatment systems or in household point-of-use (POU) systems. Centralized treatment systems that provide drinking water for a city, community, or several communities are usually attached to a distribution system [34]. Household point of- use systems are for use with on-site sources such as tube wells, which provide water to one or several households close to the facility.

Mitigating the Arsenic Problem: Social and Institutional Aspects

Awareness

Arsenic contamination is largely a natural phenomenon and their preventive measures may be public awareness campaigns, Mass media for publicizing the problem, identifying the arsenic contaminated sources, sharing arsenic free point sources and removal of arsenic at house hold level and treatment plant installed at community level.

Sharing Arsenic-Free Point Sources

Every drinking water source is not affected to arsenic due to differences in sedimentary characteristics. So, It is in principle possible to share drinking and cooking arsenic-free point source water due to socio-cultural and economic constraints.

Arsenic Removal at Household Level

A second option is the removal of arsenic at household level. Communities and households need to know about the different types of equipment that may be on the market, at what price, and how long each can be expected to last. This information needs to be available to both women and men. Demonstrations should be arranged to show operation and maintenance tasks, such as filter cleaning, and to demonstrate the effectiveness of treatment and the colour and taste of the water.

Communal Plant

The third option that the people can consider is to have a treatment plant installed at community level. A community-level solution has the advantage of being able to deliver arsenic free water to a large number of households.

Institutional Aspects

Informed choice implies good quality information, communication and decision-making processes. In the early stages, there is a need for experienced facilitators who know how to work with different user groups using participatory techniques in an equitable manner. Implementation and training can start when the operators, committee members and other functionaries have been chosen and roles and remuneration have been decided. Decisions over the design and installation of the plant and distribution net will include the location of facilities, such as a well, treatment plant, storage tank, and the distribution net to shared or individual household taps. The committee plays a central role in communicating the proposal to user groups, locating proposed sites and discussing their acceptability. Sites should be marked on a social, community-made map in which all houses; roads and key geographic features are depicted, not necessarily to scale. The location of water supply components and outlets are marked on the map and can also be marked on the ground, so that everyone can understand the new system, including ease of access to any shared distribution points and once decisions have been made, the committee will guide the implementation process.

References

1. UNICEF (2006). *Arsenic mitigation in Bangladesh* Fact Sheet. Available at: http: / /www.unicef.org/bangladesh/Arsenic.pdf
2. Smith, A.H., Lingas, E.O. and Rahman, M. (2000). Contamination of drinking-water by arsenic in Bangladesh: a public health emergency'. In: *Bulletin of the World Health Organization*, 78(9): 1093–1103.

3. Chappell, W.R., Abernathy, C.O. and Calderon, R.L. (eds) (1999). Arsenic exposure and health effects. In: *Proceeding of the Third International Conference on Arsenic Exposure and Health Effects,* 14–18 July 1998, San Diego, California, S.L., Elsevier Science.
4. Ahmed, F.M. (ed.) (2003). Treatment of arsenic contaminated water. In: *Arsenic Contamination: Bangladeshi Perspective.* Dhaka, Bangladesh, ITN–Bangladesh.
5. National Research Council (2000). *Arsenic in Drinking Water.* National Academy Press, Washington, DC, USA.
6. UN (2001). *UN Synthesis Report on Arsenic in Drinking Water.* Available at: http: //www.who.int/water_sanitation_health/dwq/arsenic3/en/
7. WHO (2001). *Arsenic in Drinking Water.* Fact sheet No. 210, Rev. ed. Available at: http: //www.who.int/mediacentre/factsheets/fs210/en/
8. Guha Mazumder, D.N. *et al.* (1998). 'Arsenic levels in drinking water and the prevalen of skin lesions in West Bengal, India'. In: *International Journal of Epidemiology,* 27(5): 871–877.
9. Rasmussen, L. and Andersen, K.J. (2002). *Environmental Health and Human Exposure Assessment.* Available: http: //www.who.int/water_sanitation_health/dwq/arsenicun2.pdf
10. IARC (2004). *Some Drinking-Water Disinfectants and Contaminants, Including Arsenic.* Lyon, France, International Agengy for Research on Cancer. (IARC monographs on the evaluation of carcinogenic risks to humans, vol. 84).
11. Morales, K.H. *et al.* (2000). 'Risk of internal cancers from arsenic in drinking water' In: *Environmental Health Perspectives,* 108(7): 655–661.
12. WHO (1993). *Guidelines for Drinking Water Quality, Vol. 1: Recommendations.* 2nd ed. Geneva, Switzerland, World Health Organization.
13. WHO (2004). *Guidelines for Drinking Water Quality, Vol. 1: Recommendations.* 3rd ed. Geneva, Switzerland, World Health Organization. Available at: http: //www.who.int/water_sanitation_health/dwq/gdwq3/en/

14. BGS and DPHE (2001). *Arsenic Contamination of Groundwater in Bangladesh, Vol. 1: Summary*. Keyworth, UK, British Geological Survey.

15. BGS and DPHE (2001b). *Arsenic Contamination of Groundwater in Bangladesh, Vol. 2: Final Report*. Keyworth, UK, British Geological Survey.

16. Csalagovits, I. (1999). Arsenic-bearing artesian waters of Hungary. In: *Annual Report of the Geological Institute of Hungary*, 1992–1993/II: 85–92.

17. Bundschuh, J., Garcia, M.E. and Birkle, P. (2006). ´Rural Latin America: a forgotten part of the global groundwater arsenic problem´. In: *Proceedings of the As 2006. International Congress: "Natural Arsenic in Groundwaters of Latin America"* (20–24 June), Mexico city, Mexico.

18. Source News (2003). *Arsenic Polluted Village in Bangladesh Loses all Hope*. IRC Source News Bulletin Features, 05 August 2003. Available at: http: //www.irc.nl/page/3044

19. Cheng, R.C., Liang, S., Wang, H.C. and Beuhler, M.D. (1994). ´Enhanced coagulation for arsenic removal´. In: *Journal AWWA*, 86(9): 79–90.

20. Ferguson, J. F. and Gavis, J. (1972). A review of the arsenic cycle in natural waters´. In: *Water Research*, 6(11): 1259–1274.

21. Hering, J.G. and Chiu, V.Q. (2000). Arsenic occurrence and speciation in municipal groundwater based supply system. In: *Journal of Environmental Engineering*, 126: 471–474.

22. Korte, N. E. and Fernando, Q. (1991). A review of arsenic (III) in groundwater. In: *Critical Reviews in Environmental Control*, 21(1): 1–39.

23. Edwards, M. *et al.* (1998). Considerations in As analysis and speciation´. In: *Journal AWWA*, 90(3): 103–113.

24. USEPA (1997). *Treatment Technology Performance and Cost Data for Remediation of Wood Preserving Sites*. Washington, DC, USA, US EPA Office of Research and Development Available at http: //www.epa.gov/nrmrl/pubs/625r97009/625r97009.pdf

25. USOSHA (2001). *Occupational Safety and Health Guidelines for Arsenic, Organic Compounds (as As)*. Washsington, DC,USA, US

Occupational Safety and Health Administration. Available at http: //www.osha.gov/SLTC/healthguidelines/arsenic/recognition.html

26. USEPA (2004). *Monitoring Arsenic in the Environment: A Review of Science and Technologies for Field Measurements and Sensors.* Washington, D C, USA, US EPA. Available at: http: //www.epa.gov/superfund/programs/aml/tech/news/asreview.htm
27. G reenberg, A.E. (ed.) (1995). *Standard Methods for the Examination of Water and Wastewater,* 19th edn. Washington, DC, USA, American Public Health Association.
28. USEPA (1999). *Analytical Methods Support Document for Arsenic in Drinking Water.* Washington, DC, USA, US EPA Office of Groundwater and Drinking Water. Available at http: //www.epa.gov/safewater/arsenic/pdfs/methods.pdf
29. Appelo, C.A.J. and Vet, W.W.J.M. de (2003). Modeling *in situ* iron removal from groundwater with trace elements such as As. In: *Arsenic in Groundwater,* (Eds.) A.H. Welch and K.G. Stollenwerk. Boston, MA, USA., Kluwer Academic. Chapter 14, p. 381–401.
30. Jacks, G., Bhattacharya, P. and Chatterjee, D. (2001). Artificial recharge as a remedial measure for *in situ* removal of arsenic from the groundwater. In: *Groundwater Arsenic Contamination in the Bengal Delta Plain Bangladesh: Proceedings of the KTH–Dhaka University Seminar,* (Eds.) Jacks, G., Bha ttacharya, P. and Khan, A.A.. Stockholm, Sweden, s.n. (KTH special publication, TRITA–AMI Report 3084).P. 71–75.
31. Katsoyiannis, I.A. and Zouboulis, A.I. (2004). ´Application of biological processes for the removal of arsenic from groundwaters´. In: *Water research,* 38(1): 17–26.
32. USEPA (2000). *Technologies and Costs for Removal of Arsenic from Drinking Water.* Washington, DC, USA, US EPA Office of Water, December, 2000. Available at http: //www.epa.gov/safewater/arsenic/pdfs/treatments_and_costs.pdf
33. USEPA (2002). *Proven Alternatives for Aboveground Treatment of Arsenic in Ground.* Washington, DC, USA, US EPA (Engineering

forum issue paper). Available at http: //www.epa.gov/tio/tsp/download/arsenic_issue_paper.pdf

34. Ahsan, T. (2002). Technologies for arsenic removal from groundwater´. In: *Small Community Water Supply,* (Eds.) J. Smet and C. van Wijk. Delft, The Netherlands, IRC International Water and Sanitation Centre (Technical paper series, no. 40).

Chapter 3

Treatment of Wastewater from Coke Oven Plant: A Case Study

D.C. Mukherjee, S.B. Chowdhury, A. Paul, B. Rakshit and Paulomi Das

ABSTRACT

The industrial progress has been accompanied by a growing negative impact on the environment in terms of its pollution and degradation. In our country number of Coke Oven batteries for production of low ash metallurgical coke is increasing day by day. During off Gas cleaning there are different sources of generation of wastewater. But the wastewater is not treated before discharge. Naturally found coal is converted into coke in coke ovens and a large quantity of water is used for quenching hot coke and for washing gas. Effluent generated contains high value of Total suspended Solid (TSS), Biological oxygen demand (BOD), Chemical oxygen demand (COD), phenols, ammonia, cyanide, which cause serious water pollution problems. This study was carried out to focus on the characteristics of the effluent produced by coke oven plant and a treatability study was conducted to find out an appropriate treatment procedure to achieve a depletion of pollutants for discharge to surface water bodies after meeting the level of standard. This study was done initially separating ammonia of raw effluent through natural zeolite for Ammonia scrubbing water. After this treatment the treated water was mixed with coal tar separator effluent, quenching effluent and gas plant effluent for chemical treatment followed by two stages Aerobic Biodegradation process and finally adsorption process by Activated Carbon. The overall depletion rate of COD, BOD, Phenol, Cyanide, in the wastewater achieved by this treatment process were 94.9 per cent, 96.2 per cent, 98.3 per cent, 34.2 per cent, respectively and the discharge effluent met the Pollution Control Board standard. On the basis of the finding the schematic treatment system was finalized.

Keywords: *Wastewater, Coke oven batteries, Chemical treatment, Biological treatment, COD, BOD.*

Introduction

Environmental pollution is one of the most serious problems facing humanity and other life forms on our planet today. In India this problem is growing rapidly. Industrial pollution, rapid industrialization, urbanization are all worsening problems. The industrial pollution due to its nature has the potential to cause irreversible reactions in the environment and hence posing a major threat to our very existence. Untreated and improperly treated waste form different industry cause pollution. The importance of coal carbonizing industries from India's national point of view is great and also growing[1,2].

In coke ovens, naturally found coal is converted into coke. Water is used in large quantities for quenching hot coke and washing of gas produced from the ovens. Thus, liquid effluents generated during the process operations are highly polluted and difficult to handle. However, coke-plant wastewaters contain ammonia, cyanide, thiocyanate and many toxic organic contaminants such as phenols, mono and polycyclic nitrogen containing compounds and PAHs[3,4]. As a result this effluent shows high value of TSS, BOD, COD[1]. Certain industries are unable to meet their permitted discharge limitations. Untreated discharge may alter the physical and chemical composition of environment. Due to this increment it is essential to find out an effective and economic treatment of wastewater for pollution control.

The aim and objectives of the study was (i) to focus on the characteristics of the effluent produced by coke oven plant (ii) to find out an appropriate treatment procedure to achieve a depletion of pollutants for discharge to surface water bodies after meeting the level of standard.

Materials and Methods

The sources of effluent for coke oven plant are Ammonia scrubbing water, Coal Tar Separator water, Quenching water, Gas Plant water. This study was done initially separating ammonia of raw effluent through natural zeolite for Ammonia scrubbing water. After this treatment the treated water was mixed with coal tar separator effluent, quenching effluent and gas plant effluent for chemical treatment followed by two stages Aerobic Biodegradation process and finally adsorption process by Activated Carbon and maturation pond.

At first, pH, TSS, COD, BOD, phenol, cyanide of mixed wastewater were measured. The instruments used in the experiment were Jar Test Apparatus, D.O. Meter, Spectrophotometer, COD Digester, pH meter, Pilot scale plant etc.

In this study the treatment process involved physical, chemical, biological and tertiary treatment process[5]. The chemicals used in treatments are given below.

Chemicals Used

1. Lime Solution (10 per cent)
2. Sodium Hydroxide Solution (5 per cent)
3. Alum Solution (10 per cent)
4. Permanganate Solution (10 per cent)
5. Urea
6. Diammonium Phosphate (DAP)
7. Activated Carbon

Principle of Treatment

Pretreatment

Separation of Ammonia from Ammonia Scrubbing Water by Natural Zeolite

The scope of this study was removal of ammonia from the scrubbing aqueous solutions by raw and pretreated natural zeolite, Transcarpathian mordenite under static and dynamic conditions. The cation exchange capacity of the Transcarpathian mordenite regarding ammonium ions was evaluated as 1.64meq/g at 1000 mg/l initial NH_4-N concentration. The dynamic exchange capacity exceeded one estimated in equilibrium study at the same initial concentration that may be conditioned by the constant removal of ion exchange products. Ammonium uptake rate was controlled by particle diffusion with diffusion coefficients determined in the range of $0.7\text{-}3.6 \times 10^{-12}\ m^2/s$. Efficiency of ammonium sorption may be improved by slowing down of initial solution rate in the column test and non-significantly by NaCl and HCl pretreatment of the mordenite.

After this pre treatment the water was mixed with coal tar separator effluent, quenching effluent and gas plant effluent for chemical treatment.

Chemical Treatment

Chemical Oxidization by $KMnO_4$

Chemical oxidization in waste treatment typically involves the use of the oxidizing agent such as Potassium permanganate to bring about change in the chemical composition of a compound or a group of compound. The use of this chemical oxidization was for the reduction of COD and BOD along with Phenol.

Half Reaction

$$MnO^-_4 + 4H^+ + 3e^- \rightarrow MnO_2 + 2H_2O$$

The oxidation of Phenol is known to take place by a stepwise mechanism. However, oxalic acid may be expected as a major final product. The stoichiometry of permanganate oxidation may thus be represented as follows.

$$8KMnO_4 + C_6H_5OH \rightarrow 8MnO_2 + 2K_2C_2O_4 + 2K_2CO_3 + 3H_2O$$

Precipitation by Lime and Alum

Here lime will be added for maintaining the alkalinity required to react with alum as well as to increase the pH value. Then alum will be added to water in the presence of alkalinity, which will produce aluminum hydroxide as gelatinous floc that will settle slowly through the wastewater and will sweep out the suspended solid as well as reduce the level of COD and BOD and finally neutralization of wastewater.

When lime alone is added as a precipitant the principles of clarification are explained by the following reactions.

$$Ca(OH)_2 + H_2CO_3 \rightarrow CaCO_3 + 2H_2O$$

$$Ca(OH)_2 + Ca(HCO_3)_2 \rightarrow 2CaCO_3 + 2H_2O$$

A sufficient quantity of lime must therefore be added to combine with all the free carbonic acid and with the carbonic acid of the bicarbonates (half bound carbonic acid) to produce calcium carbonate, which acts as the coagulant.

When Alum is added to wastewater containing calcium and magnesium bicarbonate alkalinity, the reaction that occurs may be illustrates as below.

$$Al_2(SO_4)_3 . 18H_2O + 3Ca(HCO_3)_2 \rightarrow 3CaSO_4 + 2Al(OH)_3 + 6CO_2 + 18H_2O$$

The insoluble aluminum hydroxide is a gelatinous floc that settles slowly through the wastewater, sweeping out suspended material and producing other changes.

In different experiments with different composition of various chemicals was tried for Chemical treatment.

In the bench scale type 1 lit of mixed effluent water was taken and permanganate solution was added to keep for oxidation reaction. After reaction and settling the supernatant solution was collected and Lime and Alum solution was added to bring the pH around 7.5-8.0.

Biodegradation

The chemically treated effluent water was taken in different types of Biodegradation process.

Aerobic Treatment

A bench scale batch type aerobic reactor was used for carrying out the study of treatment taking the chemically treated effluent. Specific aerobic stain was developed and used in the system. The system was provided with compressed air as per requirement round the clock. The sample was collected time to time to see the per cent degradation of organic load. MLSS was maintained at 2000-3000 mg/lit. The Aerobic treatment was done in two stages and found that 48 hrs. of retention time in 1st stage and 4 days in 2nd stage will be more appropriate one to bring down the BOD level below 30 mg/lit.

Tertiary Treatment

In tertiary treatment the clear effluent from aerobic treatment was passed through a column of activated carbon. After activated carbon column the clear effluent was collected and analyzed. Two activated carbon columns in series was suitable in achieving the best result in the tertiary system.

Tertiary maturation or low rate stabilization ponds may be constructed to provide for secondary effluent polishing and seasonal nitrification. The biological mechanisms involved are similar to other aerobic suspended growth processes. Operationally, the residual biological solids are endogenously respired and ammonia is converted to nitrate using the oxygen supplied from surface reaeration and from algae. A detention time of 18 to20 days has been found as the minimum period required to provide for complete endogenous respiration of the residual solids.

Systems Description

The effluent water generated mainly from Ammonia scrubbing section, Coal tar separation system, Quenching and Gas Plant unit. At first Ammonia scrubbing water passing through natural zeolite bed for Ammonia separation. Then zeolite treated water mixed with another three water generated from Coal tar, Quenching and Gas Plant section.

The mixed water will be first brought to an equalization tank. From equalization tank the effluent will be lifted to Continuous Stirred Tank No. 1 (CSTR-1) where Potassium Permanganate ($KMnO_4$) solution be added to oxidize the organic portion of the effluent. The effluent passed to Primary Settling tank No. 1 for proper precipitation. The over flow from Primary Settling tank No. 1 will be brought to CSTR No. 2 where Lime solution will be added to increase the alkalinity. The overflow will be brought to CSTR No. 3 where appropriate quantity of Alum solution will be dosed for coagulation. The overflow from CSTR No. 3 will then brought to Primary settling tank-2 where the precipitation of the coagulated mass will occur. The precipitation will be settled at the bottom of the tank as sludge which will be brought to sludge drying bed by gravity.

The clear liquid from Primary Settling tank-2will be brought to Aeration tank No. 1 where Primary Aerobic Biodegradation of Biodegradable Organic part of the effluent will occur. The overflow from Aeration tank No. 1 will be brought to Secondary settling tank/ clarifier No. 1 where biological mass will be settled at the bottom. A part of the sludge will be recycled (by a sludge pump) back to the Aeration tank No. 1 for recycling and other part will be brought to sludge drying bed. The clear effluent from the Secondary Settling tank No. 1 will be brought to Aeration tank No. 2 where second stage Aerobic Biodegradation will occur. The overflow from Aeration tank No. 2 will be brought to Secondary Settling tank No. 2, where biological mass will be settled at the bottom. A part of the Biological sludge will be back to the Aeration tank No. 2 to maintain the MLSS of the effluent at desired level. The other part will be brought to sludge drying bed. The clear liquid from Secondary Settling tank No. 2 will pass through two activated carbon beds in series and finally discharge to Maturation pond. After Maturation pond the treated water will be recycled or discharge.

Results and Discussion

The manufacturing processes for coke generate considerable quantities of wastewater to be discharged to the sewer and irrigation channel. These effluents are characterized by high COD and BOD, Cyanide, Phenol[6]. Table 3.1 shows the water quality of the mixed waste used for the treatment. Different dosing of Chemicals are showed in Table 3.2. Changes in the process parameters after treatment are depicted in Table 3.3. (per cent) Depletion of pollutants in different steps are given in Table 3.4.

Table 3.1: Characteristics of Mixed Effluent Before Treatment from Coke Oven Plant

pH	9.78
TSS (mg/l)	214.5
COD (mg/l)	2199
BOD (mg/l)	675
Phenol (mg/l)	60
Cyanide (mg/l)	12.8

Table 3.2: Different Dosing of Chemicals

Exp. No.	Dosing Chemicals	Reduction in (per cent)				
		TSS	COD	BOD	Phenol	Cyanide
1	5 ml $KMnO_4$ soln. (10 per cent) + 1 ml Lime (10 per cent) + 5 ml Alum (10 per cent)	42	23	20	45	–
2	10 ml $KMnO_4$ soln. (10 per cent) + 2 ml Lime (10 per cent) + 8 ml Alum (10 per cent)	48	30	26.5	65	1.56
3	12 ml $KMnO_4$ soln. (10 per cent) + 3 ml Lime (10 per cent) + 10 ml Alum (10 per cent)	53	32.5	31	80	1.56
4	15 ml $KMnO_4$ soln. (10 per cent) + 4 ml Lime (10 per cent) + 12 ml Alum (10 per cent)	58	36.88	33.33	91.1	2.34

*Treatment: 1 Lit of sample + 15 ml. 10 per cent of $KMnO_4$ (settling time 2 hrs.).. Supernatant + 4 ml. 10 per cent lime + 12 ml. 10 per cent alum. (settling time 30 mins.).... Supernatant for Biological treatment.

Table 3.3: Characteristics of Mixed Effluent After Treatability Study by Chemical and Microbiological Treatment

	Water Sample	*pH*	*TSS (mg/l)*	*Phenol (mg/l)*	*Cyanide (mg/l)*	*COD (mg/l)*	*BOD (mg/l)*
A.	After Chemical Treatment*	8.2	90	5.35	12.5	1388	450
B.	After 1st Biological Treatment						
i	After 24 hrs.	7.8	65	4.5	9	1148	338
ii	After 36 hrs.	7.6	58	3.2	6.5	825	195
iii	After 48 hrs.	7.3	45	1	3	540	120
C.	After 2nd Biological Treatment						
i	After 72 hrs.	7.3	38	BDL	1	285	68
ii	After 96 hrs.	7.2	35	BDL	BDL	165	40
D.	After Activated Carbon	7.2	25	BDL	BDL	135	35
E.	After maturation pond (expected results)	7.3	42	BDL	BDL	110	25

*Treatment: 1 Lit sample + 15 ml. 10 per cent of $KMnO_4$ (settling time 2 hrs.). Supernatant + 4 ml. 10 per cent lime + 12 ml. 10 per cent alum. (settling time 30 mins.).... Supernatant for Biological treatment.

Table 3.4: (Per cent) Depletion of Pollutants in Different Steps of Treatment

	Water Sample	*TSS (per cent)*	*Phenol (per cent)*	*Cyanide (per cent)*	*COD (per cent)*	*BOD (per cent)*
A	After Chemical Treatment*	85	91.1	2.4	36.88	33.33
B	After 1st Biological Treatment (after 48 hrs.)	21	7.2	76.5	38.56	48.8
C	After 2nd Biological Treatment (after 96 hrs.)	4.6	–	15.62	17.05	11.85
D	After Activated Carbon	4.6	–	–	1.36	0.74
E	After Maturation Pond (expected degradation)	(-)7.9	–	–	1.13	1.48
		80.3	98.3	94.52	94.9	96.2

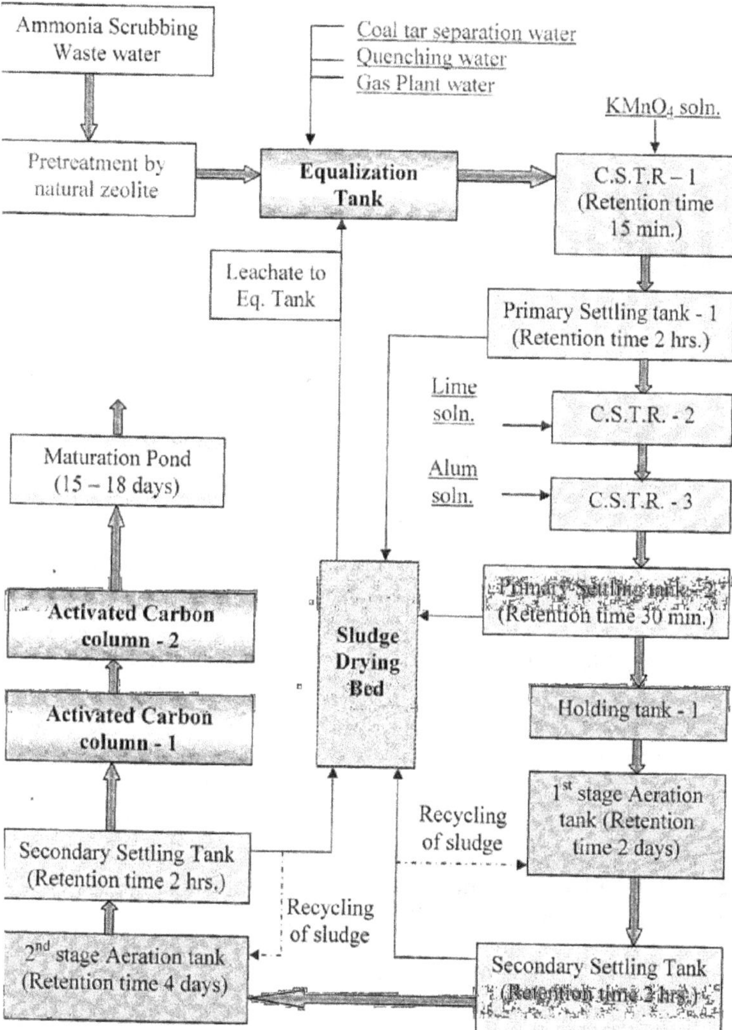

Figure 3.1: Schematic Diagram of Treatment Plant for Leather Chemical Effluent

Chemical Treatment

The system included wastewater neutralization, pH adjustment. The goals of the system were neutralization, pH adjustment and to reduce suspended solid as well the level of COD and BOD.

Physico-chemical treatment can be considered as a suitable option for the treatment of coke plant effluent[1]. In the present study an attempt had been made first to remove ammonia from the scrubbing aqueous solutions by raw and pretreated natural zeolite. This pretreatment was effective to remove the ammonia and similar to study by Qasim[7] where wastewater passed through a bed of clinoptilolite (a zeolite resin), which selectively removed the ammonium ion. Ghose[8] suggested a physico-chemical mode of treatment for coke plant effluents by addition of lime slurry in a secondary sedimentation tank, while efficient ammonia removal might be achieved by synthetic zeolite columns.

The conventional process used to treat coke oven wastewater includes chemical neutralization[9]. For mixed effluent Potassium permanganate was applied as oxidizing agent such as to bring about change in the chemical composition of the liquid followed by chemical coagulant alum in the presence of alkali, as a pretreatment which produced aluminum hydroxide as gelatinous floc that settled slowly through the wastewater and sweeped out the suspended solid. From Table 3.2, it was shown that among different dosing 15ml $KMnO_4$ soln. (10 per cent) along with 4 ml Lime (10 per cent) and 12 ml Alum (10 per cent) was best to achieve the maximum depletion. The wastewater with a TSS value of 214.5 mg/L was reduced to 90 mg/L. 58.0 per cent depletion of TSS value was observed in this step which helped in increasing the efficiency for achieving better biological treatment[10]. This study was also similar to the study of Kabdasli *et al.*[11]. Throop[12] also reported that potassium permanganate was effective to reduce phenol content in wastewater. The initial COD, BOD and phenol were also reduced by 36.88 per cent, 33.33 per cent and 91.1 per cent respectively.

Aerobic Treatment

Biological treatment may be considered as a suitable option for the treatment of coke plant effluents and control of water pollution. Prasad and Singh[6] reviewed different treatments such as chemical and biological processes for coke oven effluent and found that aerobic biological treatment are more common in practice. To start the operation, diluted mixed wastewater after chemical treatment with COD of 1388 mg/L was introduced. Appreciable COD and BOD removal did not occur within the first 24 hours of micro biological treatment. But in case of after 48 hours, COD and BOD

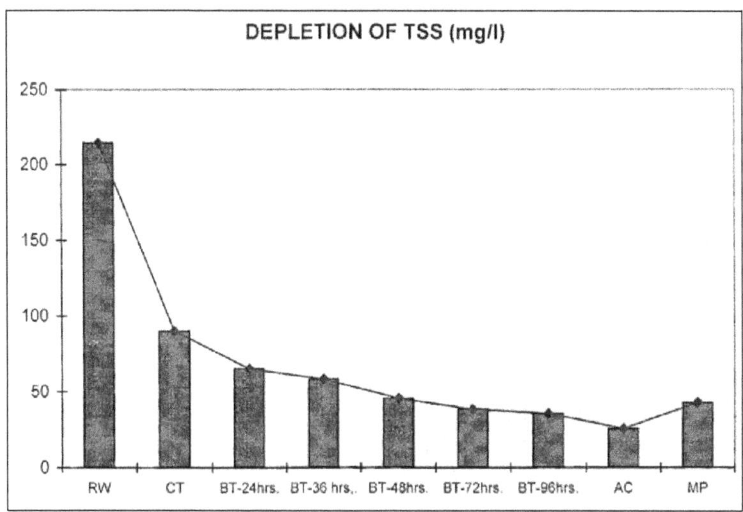

Figure 3.2: Showing (per cent) Depletion of TSS in Different Steps of Treatment

RW=Raw Water; CT=Chemical Treatment; BT=Biological Treatment; AC=Activated carbon, MP=Maturation Pond. Limit-100 mg/l

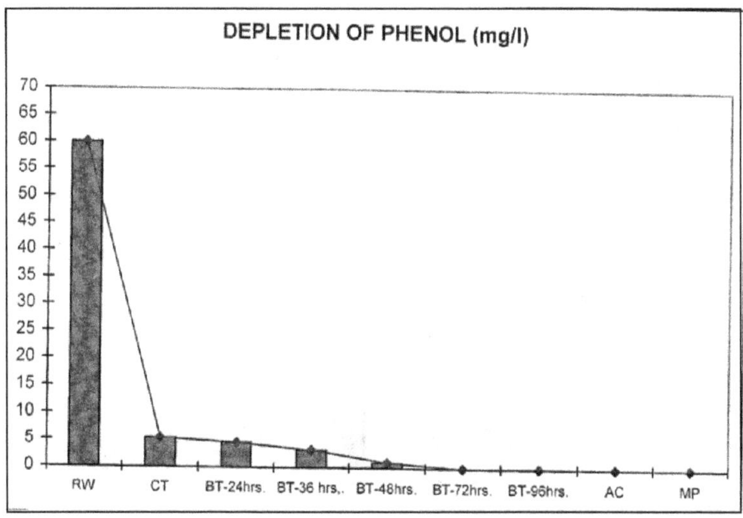

Figure 3.3: Showing (per cent) Depletion of Phenol in Different Steps of Treatment

RW=Raw Water; CT=Chemical Treatment; BT=Biological Treatment; AC=Activated carbon, MP=Maturation Pond. Limit-0.05 mg/l

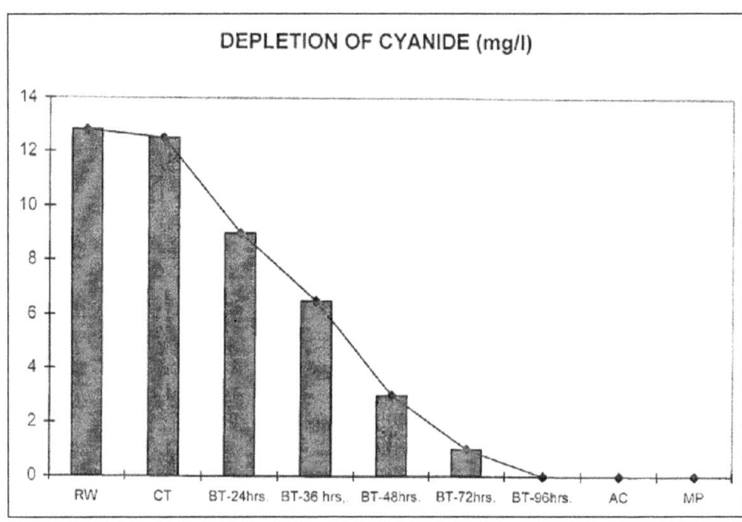

Figure 3.4: Showing (per cent) Depletion of Cyanide in Different Steps of Treatment

RW=Raw Water; CT=Chemical Treatment; BT=Biological Treatment; AC=Activated carbon, MP=Maturation Pond. Limit-0.02 mg/l

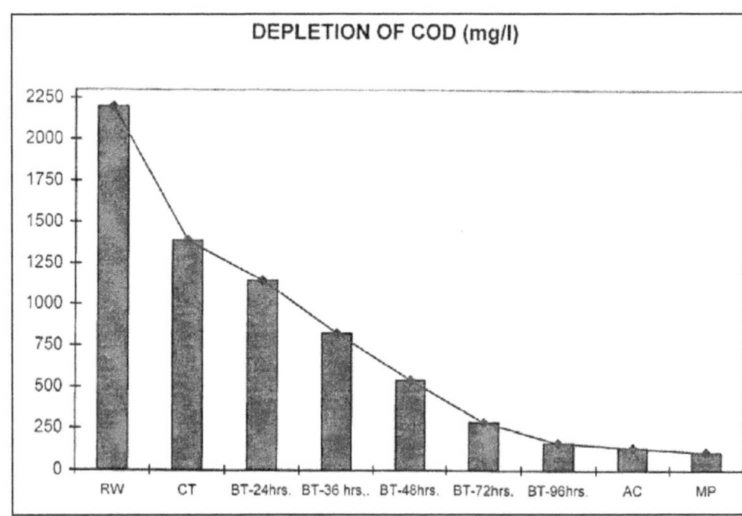

Figure 3.5: Showing (per cent) Depletion of COD in Different Steps of Treatment

RW=Raw Water; CT=Chemical Treatment; BT=Biological Treatment; AC=Activated carbon, MP=Maturation Pond. Limit-250 mg/l

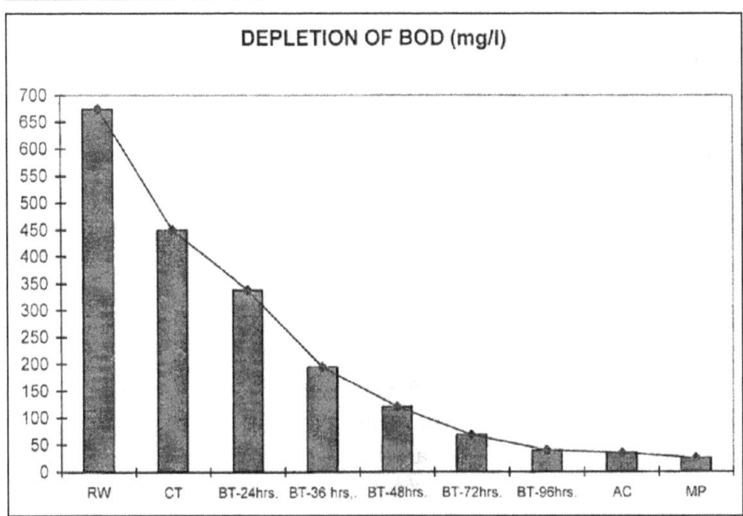

Figure 3.6: Showing (per cent) Depletion of BOD in Different Steps of Treatment

RW=Raw Water; CT=Chemical Treatment; BT=Biological Treatment; AC=Activated Carbon, MP=Maturation Pond. Limit-30 mg/l

were 540 mg/L and 120 mg/L respectively. Maximum reduction of COD and BOD after 96 hours was observed. Maximum depletion of 76.5 per cent was also noticed after 48 hrs. of biological treatment in case of cyanide whereas it showed BDL value after 96 hrs. Phenolic compounds and organic trace contaminants are biodegradable and hence aerobic biological treatment are most applicable treatment for coke-oven effluent[6]. In our study phenol showed BDL value after 96 hours of biological Treatment. Biological removal of cyanide compound was also reported by Sirianuntapiboon *et al.*[13].

Tertiary Treatment

Sakoda *et al.*,[14] developed an activated carbon membrane to be used in water treatments and successfully demonstrated the decolorization of the coke furnace wastewater as a model case. Activated carbon adsorption has the BOD, COD removal efficiencies[15] and can be used to remove the color, odor from wastewater. Zhang *et al.*[16] showed enhanced carbon adsorption treatment for removing cyanide from coking plant effluent. In our present study, TSS, COD and BOD depletion of coke oven effluent obtained with activated carbon treatment were about 4.6, 1.36 and 0.74 percent respectively where as Cyanide showed BDL value before introducing activated carbon. Johnson *et al.*[17] reported the ability of maturation pond. In coke oven effluent, after maturation pond, percent depletion for COD and BOD were 1.13and 1.48 respectively.

Over all depletion of TSS, phenol, cyanide, COD and BOD after chemical, biological and tertiary treatment were 80.3, 98.3, 94.52, 94.9 and 96.2 percent respectively.

Conclusions

Coke plant effluents are one of the contributors to the pollution. Appropriate treatment techniques including Chemical and Biological treatments for coke oven effluent have been conducted due to their cost-effectiveness and environmental acceptability. From the above experiments, the best treatment process was finalized and these treatments were able to deplete the pollution load to below the permissible limit of Pollution Control Board.

So, from the study it might be concluded that chemical treatment followed by biological and tertiary treatments was the best technique for controlling pollution due to coke oven effluents. By adopting the suggested treatment scheme an effective treatment plant for coke plant effluent may be designed and the treated effluent can be recycled or safely discharged to water bodies.

Acknowledgement

The authors gratefully acknowledge the cooperation of Durgapur Projects Ltd., Durgapur Burdwan, West Bengal to carry out this study.

References

1. Ghose, M.K., Bhasa, S.K. and Jena, A. (2006). *Indian Chemical Engineer.*, 48(4): 278–287.
2. Ghose, M.K. and Roy, S. (1999). *J. of the Air and Waste Management Association*, 49: 1245–1249.
3. Suschka, J., Morel, J., Mierzwinski, S. and Janusznek, R. (1994). *Wat. Sci. Tech.*, 29: 69–79.
4. Li, Y.M., Gu, G.W., Zhao, J.F., Yu, H.Q., Qui, Y.L. and Peng, Y.Z. (2003). *Chemosphere*, 52: 997–1005.

5. Arceivala, S.J. and Asolekar, S.R. (2008). *Wastewater Treatment for Pollution Control and Reuse,* 3rd Edn. Tata McGraw-Hill Professional, New Delhi.
6. Prasad, B. and Singh, G. (1989). *Indian Journal Environmental Protection,* (7): 525–530.
7. Qasim, S.R. (1985). *Wastewater Treatment Plant: Planing, Design and Operation.* Holt, Richart and Winston.
8. Ghos, M.K. (2002). *Water Research,* 36: 1127–1134.
9. Chang, E.-E., Hsing, H.-J., Chiang, P.-C. and Shyng, M.-Y. (2008). *Journal of Hazardous Materials,* 156: 560–567.
10. Wang, K., Liu, S., Zhang, Q. and He, Y. (2009). *Environ Technol.,* 30(13): 1469–74.
11. Kabdasli, I., Gürel, M. and Tünay, O. (1999). *Water Science and Technology,* 39(10–11): 265–271.
12. Throop, W.M. (1975-1976). *Journal of Hazardous Materials.* 1(4): 319–329.
13. Sirianuntapiboon, S., Chairattanawan, K. and Rarunroeng, M. (2008). *J. Hazard Mater.,* 154(1–3): 526–534.
14. Sakoda, A., Nomura, T. and Suzuki, M. (1996). *Adsorption,* 3(1): 93–98.
15. Malik, L., Al-Hashimi, M.A. and Al-Doori, M.M. (2007). *Desalination,* 216(1–3): 116–122.
16. Zhang, W., Liu, W., Lv, Y., Li, B. and Ying, W. (2010). *J. Hazard Mater.,* 184(1–3): 135–40.
17. Johnson, M., Camargo Valero, M.A. and Mara, D.D. (2007). *Water Science and Technology,* 55(11): 135.

Chapter 4

Solar Photocatalytic Technique for the Water Purification

Subrata Saha and Anil Kumar

ABSTRACT

Solar energy is the most common eco-friendly and economical energy sources. It is effective in removing salts/ minerals (Na, Ca, As, F, Fe, Mn), bacteria (*E. coli*, Cholera, botulinus), parasites (Giardia, Cryptosporidium) heavy metals (Pb, Cd, Hg) and also distilled seawater even raw sewage for drinking water, household purposes, charging the batteries and medical appliances etc. These are the following processes like solar photocatalytic system, solar pasteurization, Hybrid solar water purification, solar photovoltaic system, Plastic solar water purifier etc based on the solar energy for solar water purification. A photo catalytic process along with titanium dioxide, based on UV solar radiation provides an excellent route to destroy bacteria and reduce significant amount of organic hazardous contaminants. Removal of α-lindane (99 per cent), ethyl benzene and other toxic compound from water is commercially successful and effective in this photo catalytic process. This Chapter focuses on the removal of contaminants present in water through photocatalytic technique.

Keywords: *Photocatalytic process, Solar radiation, Toxic compounds, Titanium dioxide and Drinking water treatment.*

Introduction

Increasing demand and shortage of clean water sources due to the rapid development of industrialisation, population growth and long-term droughts have become an issue worldwide. With growing demand, a various techniques and solutions have been adopted to minimise this problem. A photo catalytic detoxification process has been discussed as an alternative method for clean-up polluted water in the scientific literature since 1976 [1]. Titanium dioxide has been demonstrated to be photo catalytically active in the presence of sunlight. Although titanium dioxide utilizes only 3 per cent to 4 per cent of the energy in the solar UV spectrum (300-400nm), it is the most widely used photo catalyst for organic photo destruction due to its high catalytic activity, stability in acidic and basic media and no toxicity. Organochlorine pesticides such as lindane can also remove by this process. Considerable public attention has been focused on this possibility of combining heterogeneous catalysis with solar technologies to achieve the mineralization of toxins present in water [2]. Compilations of substances which can mineralized using photo catalysis are available [3]. Several reviews have been published discussing the underlying reaction mechanisms and illustrating examples of successful laboratory and field studies [4].

Mechanisms of Photocatalysis

Semiconductors (*e.g.*, TiO_2, ZnO, Fe_2O_3, CdS and ZnS) can act as sensitizers for light-induced redox processes due to their electronic structure which is characterized by a filled valence band and an empty conduction band. Absorption of a photon of energy greater than the band gap energy leads to the formation of an electron/hole pair. In the absence of suitable scavengers, the stored energy is dissipated within a few nanoseconds by recombination [5]. Most organic photo degradation reactions utilize the oxidizing power of the holes either directly or indirectly however, to prevent a buildup of charge one must also provide a reducible species to react with the electrons. In bulk semiconductor electrodes only one species, either the hole or electron, is available for reaction due to band bending [6]. However, in very small semiconductor particle suspensions both species are present on the surface. Therefore, careful consideration of both the oxidative and the reductive paths is required.

On the surface of the semiconductor particles following the absorption of a photon is yet unclear. A significant body of literature exists that the initial oxidation of a pollutant molecule may either occur by indirect oxidation via a surface-bound hydroxyl radical. It is shown in Figure 4.1. It consists of a superposition of the energy bands of a generic semiconductor (valence band VB, conduction band CB) and the geometrical image of a spherical particle. Absorption of a photon with an energy hn greater or equal the band gap energy is generally leads to the formation of an electron/hole pair in the semiconductor particle. These charge carriers subsequently either recombine and dissipate the input energy as heat and get trapped in metastable surface states or react with electron donors and acceptors adsorbed on the surface or bound within the electrical double layer. The exact nature of the main oxidizing species formed surface), directly via the valence band hole before it is trapped either within the particle or at the particle surface or even via hydroxyl radicals in solution. In support of a hydroxyl radical as the principal reactive oxidant in photo activated TiO_2 is the observation that

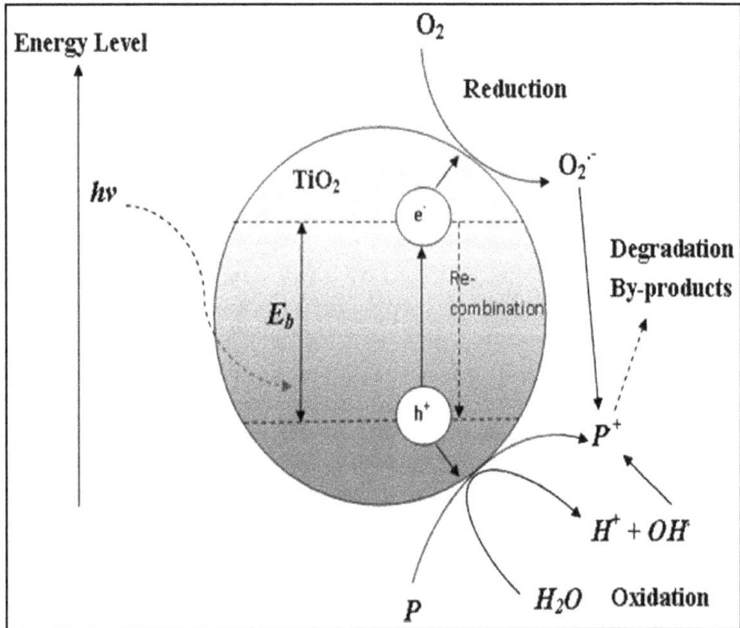

Figure 4.1: Photo-Induced Formation Mechanism of Electron Hole Pair in a Semiconductor

intermediates detected during the photo catalytic degradation of halogenated aromatic compounds are typically hydroxylated structures [7]. These intermediates are consistent with those found when similar aromatics are reacted with a known source of hydroxyl radicals. In addition, ESR studies have verified the existence of hydroxyl and hydroperoxyl radicals in aqueous solutions of illuminated TiO_2 [8]. The strong correlation between degradation rates and concentration of the organic pollutant adsorbed to the surface also implies that the hydroxyl radicals are adsorbed at the surface [9]. It should be noted that these compounds also have no hydrogen atoms available for abstraction by OH^-.

The primary processes occurring upon band gap irradiation of extremely small titanium dioxide particles (diameter d=2.5nm) have been studied extensively employing ultra-fast laser flash photolysis equipment with Picosecond or even sub picosecond time-resolution [10]. While quantitative discrepancies were present in different preparation techniques of the TiO_2 colloids and the authors were agreed principally on the underlying qualitative concepts of photo catalysis. The trapped electron [11] exhibits strong transient optical absorption around 650nm while the trapped hole absorbs predominantly at shorter wavelengths (430nm). Studying the decay kinetics of this type of transient absorption spectra in the presence of molecular oxygen, air or molecular nitrogen and identified that the hydrated electrons possess a considerable driving force for the reduction of molecular oxygen to form the superoxide radical O_2 at pH 3 the conduction band electrons of the colloidal TiO_2 particles barely reach a one-electron potential enough to reduce O_2 [12-13]. Photocatalytic detoxification process consists of reductive and oxidative paths for the degradation of various types of pollutants [14 –19].

Solar Reactors and Pilot Plants for Water Purification

The Sun light is the economical and ecological source as photons for the detoxification of polluted water. Principally these photons are responsible for destroying the present the microorganisms in wastewater. On the basis of above mechanism following reactors are present now a days.

Parabolic Trough Reactor (PTR)

A parabolic trough reactor (PTR) concentrates the parallel (direct) rays of the photo catalytically active ultra-violet part of the solar spectrum. It is characterized as a typical plug flow reactor [20-21]. The Figure 4.2 Shows a schematic flow chart of the parabolic trough reactor (PTR). The solar rays were concentrated by parabolic trough mirrors (total aperture area: 192 m^2) with an aluminized UV-reflective surface and focused on borosilicate glass tubes, which were filled with the contaminated aqueous TiO_2 suspensions moving at flow rates between 250 and 3500 Lh^{-1}. Due to losses caused by reflectivity, translucence and system errors the yield of the UV-light photons reaching the contaminated suspensions is about 58 per cent of the original light intensity entering the aperture plane.

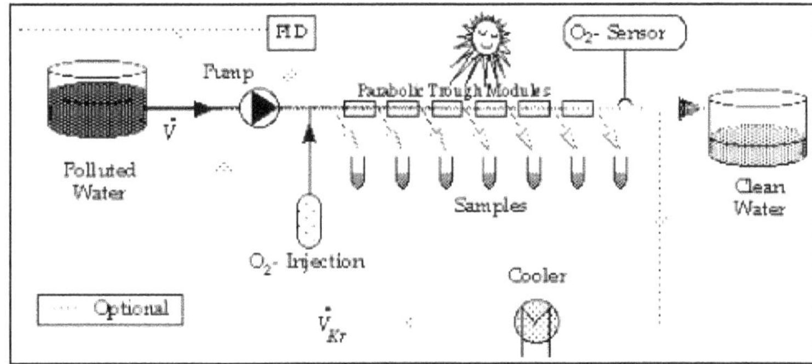

Figure 4.2: Flow Chart of PTR Installation Located at the Platform Solar de Almeria in Spain (Adopted from Sanchez, 1993)

Thin Film Fixed Bed Reactor (TFFBR)

This reactor utilized the diffusion irradiation of solar system for photocatalytic process. It is shown in Figure 4.3. The most important part of the thin-film-fixed-bed reactor has a sloping plate (width 0.6m × height 1.2m) coated with the photo catalyst and rinsed with the polluted water in a very thin film (100μm). The flow rate is controlled by a cassette peristaltic pump and can be varied between 1 and 6.5Lh^{-1}

Compound Parabolic Collecting Reactor (CPCR)

A compound parabolic collecting reactor (CPCR) is a trough reactor without light concentrating properties. It differs from a

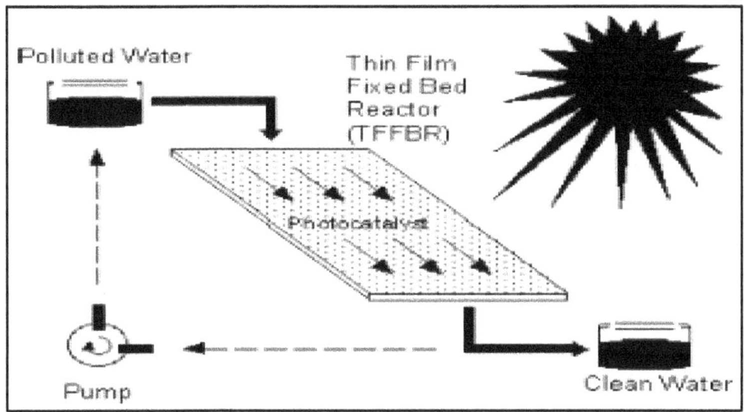

Figure 4.3: Flow Chart of a TFFBR Reactor (Adopted from Goslich *et al.*, 1997)

conventional parabolic trough reactor by the shape of its reflecting mirrors [22]. It is shown in Figure 4.4.

Double Skin Sheet Reactor (DSSR)

A new kind of non-concentrating reactor is the double skin sheet reactor (DSSR). It consists of a flat and transparent structured box made of PLEXIGLAS [23-24]. This type of reactor can utilize both the direct and the diffuse portion of the solar radiation in

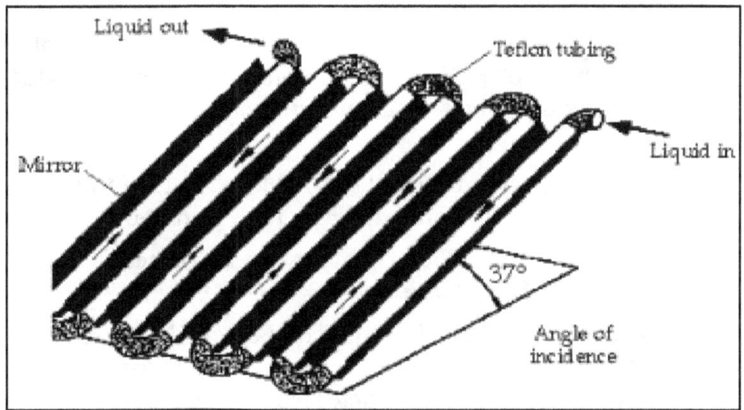

Figure 4.4: Schematic View of One CPCR-Module Installed at the Platform Solar de Almeria in Spain (Adapted from Malato *et al.* (2002)

analogy to the CPCR. After the degradation process the photocatalyst has to be removed from the suspension either by filtering or by sedimentation for both reactors.

Advantages

- ☆ Semiconductor photocatalytic technology using solar light for disinfectants and mineralization.
- ☆ It is important as recalcitrant organics
- ☆ Different water contaminants like pesticides, herbicides and detergents to pathogens, viruses, coliforms and sporesare effectively removed by this photocatalytic process.
- ☆ The residence time and reaction volume for the biological treatment could be significantly reduced.
- ☆ Photocatalytic process is used as a stand-alone treatment system, the residence time required might be prolonged for total bacterial inactivation or mineralization.

Future Challenges

The applicability of the heterogeneous photocatalytic technology for water treatment is constrained by several key technical issues that need to be further investigated.

References

1. Carey, J.H., Lawrence, J. and Tosine, H.M. (1976). Photo-dechlorination of PCBs in the presence of titanium dioxide inaqueous suspensions. *Bull. Environ. Contam. Toxicol.*, 16: 697–701.
2. Legrini, O., Oliveros, E. and Braun, A.M. (1993). Photochemical processes for water treatment. *Chem. Rev.*, 93: 671–698.
3. Kamat, P.V. (1993). Photochemistry on non-reactive (Semi-conductor) surfaces. *Chem. Rev.*, 93: 267–300.
4. Rothenberger, G., Moser, J., Gratzel, M., Serpone, N. and Sharma, D.K. (1985). Charge carrier trapping and recombination dynamics in small semiconductor particles. *J. Am. Chem. Soc.*, 107: 8054–8059.

5. Memming, R. (1988). In: Photoelectrochemical solar energy conversion. In: *Topics in Current Chemistry*, (Ed.) E. Steckham Springer, Berlin, Heidelberg, New York, 143: 79–111.

6. Augugliaro, V., Palmisano, L., Sclafani, A., Minero, C. and Pelizzetti, E., 1988. Photocatalytic degradation of phenol in aqueous titanium dioxide dispersions. *Toxicol. Environ. Chem.*, 16: 89–109.

7. Noda, H., Oikawa, K. and Kamada, H. (1993). ESR spin-trapping study of active oxygen radicals from photoexcited semiconductors in aqueous H_2O_2 solutions. *Chem. Soc. Jpn.*, 66: 455–458.

8. Ohtani, B. and Nishimoto, S. (1993). Effect of surface adsorptions of aliphatic alcohols and silver ion on the photocatalytic activity of TiO_2 suspended in aqueous solutions. *J. Phys. Chem.*, 97: 920–926.

9. Serpone, N., Lawless, D., Khairutdinov, R. amd Pelizzetti, E. (1995). Subnanosecond relaxation dynamics in TiO_2 colloidal sols (particle sizes R(P) = 1.0–13.4nm)–relevance to heterogenous photocatalysis. *J. Phys. Chem.*, 99: 16655–16661.

10. Bahnemann, D.W., Henglein, A., Lilie, J. and Spanhel, L. (1984a). Flash photolysis observation of the absorption spectra of trapped positive holes and electrons in colloidal TiO_2. *J. Phys. Chem.*, 88: 703–711.

11. Serpone, N. and Pelizzetti, E. (1989). *Photocatalysis: Fundamentals and Applications*. Wiley, New York.

12. Duonghong, D., Ramsden, J. and Gratzel, M. (1982). Dynamics of interfacial electron-transfer processes in colloidal semiconductor systems. *J. Am. Chem. Soc.*, 104: 2977–2985.

13. Serpone, N. and Pelizzetti, E. (1989). *Photocatalysis: Fundamentals and Applications*. Wiley, New York.

14. Hilgendorff, M., Hilgendorff, M. and Bahnemann, D.W. (1996). Mechanisms of photocatalysis the reductive degradation of tetrachloromethane in aqueous titanium dioxide suspensions. *J. Adv. Oxid. Technol.*, 1: 35–43.

15. Blake, D.M. (1994). Bibliography of work on the photocatalytic removal of hazardous compounds from water and air. Report

NREL/TP–430–6048, National Renewable Energy Laboratory, USA.

16. Legrini, O., Oliveros, E. and Braun, A.M. (1993). Photochemical processes for water treatment. *Chem. Rev.*, 93: 671– 698.
17. Oudjehani, K. and Boule, P. (1992). Photoreactivity of 4-chlorophenol in aqueous solution. *J. Photochem. Photobiol.* A 68: 363–373.
18. Al-Sayyed, G., D-Oliveira, J.C. and Pichat, P. (1991). Semiconductor-sensitized photodegradation of 4-chlorophenol in water. *J. Photochem. Photobiol.*, A 58: 99–114.
19. Crittenden, J.C., Zhang, Y., Hand, D.W., Perram, D.L. and Marchand, E.G. (1996). Solar detoxification of fuel-contaminated groundwater using fixed-bed photocatalysts. *Water Environ. Res.*, 68: 270–278.
20. Alpert, D.J., Sprung, J.L., Pacheco, J.E., Prairie, M.R., Reilly, T.A., Milne, T.A. and Nimlos, M.R. (1991). Sandia national laboratories work in solar detoxification of hazardous wastes. *Solar Energy Mater.*, 25: 594–607.
21. Okamoto, K., Yamamoto, Y., Tanaka, H. and Itaya, A. (1985). Heterogeneous photocatalytic decomposition of phenol over TiO_2 powder. *Bull. Chem. Soc. Jpn.*, 58: 2015–2022.
22. Malato, S., Blanco, J., Vidal, A. and Richter, C. (2002). Photocatalysis with solar energy at a pilot-plant scale: An overview. *Appl. Catal. B–Environ.*, 37: 1–15.
23. Goswami, D.Y., Klausner, J.F., Mathur, G.D., Martin, A., Wyness, P., Schanze, K., Turchi, C. and Marchand, E. (1993). Solar photocatalytic treatment of groundwater at Tyndall AFB: Field test results. In: *Proceedings of the 1993 Annual Conference of the American Solar Energy Society*, pp. 235–239.
24. van Well, M., Dillert, R.H.G., Bahnemann, D.W. and Benz, V.W., Muller, M.A. (1997). A novel non-concentrating reactor for solar water detoxification. *Trans. ASME, J. Solar Energy Eng.*, 119: 114–119.

Chapter 5

Chemistry of the Health Hazards of Fluoride Poisoning in India

M.N. Majumder

Introduction

Fluoride is a dangerous inorganic chemical poison, impacting man, animals and plants, which in India is mostly geogenic, that is, of geological origin, carried to the surface mainly through tubewell water. If the groundwater flows over or stay in contact with fluoride minerals (like fluorite, fluorapatite, biotite, amphibole, hornblende, mica) then some fluoride enters the water because water is known to be almost an universal solvent. This waterborne fluoride comes in small concentrations *viz.* from 0.1 to even 48 mg/litre and when in water imparts no color, odor, taste or smell and hence cannot be suspected through our senses. Only through sophisticated chemical analyses their presence can be known and concentrations determined. The diseases caused through ingestion of fluoride is called fluorosis like arsenicosis. Fluoride is almost as harmful as arsenic that is now ruining wide areas of W.B., the Gangetic plains and Bangladesh. The biological effects of fluoride is more severe than that of lead.

India at present is the most fluoride afflicted country in the world and fluoride poisoning is very high in most areas of Jharkhand, while that in Rajasthan, Gujarat, AP, Chhattisgarh, Bihar, UP are higher still. At present 224 districts of 19 Indian states are badly affected and the people affected by fluorosis is more than two and a half crore while more than 7 crores of Indians are at risk of fluorosis. People in the risk zones are constrained to drink water containing more than 1.5 mg/litre, which is the WHO accepted upper fluoride permissible limit. The actual number of people harbouring low-level chronic fluoride toxicosis (*i.e.* poisoning) symptoms must be much more. Good nutrition along with adequate intake of various vitamins, anti-oxidants, minerals and green vegetables partly neutralize the poisoning effects of fluoride. But the poor rural people neither have the supply of fluoride-safe drinking water, nor they have the means for wholesome nourishment. So they suffer most. The tribal peoples' sufferings are most severe.

The Government of India accepts 1.0 mg/litre as the upper permissible limit for fluoride in drinking water. But it must not be construed that 1.0 mg/litre is safe. Recent scientific researches have revealed that *fluoride in any concentration in living systems is harmful.* Less fluoride less harm, more fluoride more harm. Again smaller concentrations of fluoride will cause much more adverse health effects in a malnourished person. Fixing of 1.0 mg or 1.5 mg/litre as the permissible upper limit is due to the technological difficulties of measurement and involved costs. Even much less than 1.0 mg/litre (say 0.3 to 0.6) have been found to have deleterious effects on man and animals.

Chemical Pollution and Ruin of Civilizations

There are examples in history where environmental degradation/chemical pollution ruined some civilizations which may be repeated even now if we do not care to understand and remember lesson from history. The Sumerian, Babylonian, Harappan civilizations were ruined through soil salinization. Lead poisoning contributed to the decline and fall of Roman civilization. Palmyran (near Damascas) civilization became ruined through fluoride poisoning. This is revealed through the paleopathological studies of the teeth and skeletal remains from the tombs of the south-east necropolis of Palmyra by Japanese archaeologists in the 1990s, where there was a flourishing civilization some 2000 years ago during the Roman times.

The Strange Controversy and Obscurity that Still Surrounds the Biological Effects of Fluoride

The poisonous history of arsenic spans over 25 centuries. Arsenic was used both as a medicine and poison for so many centuries until at last only in the second half of the 20th century arsenic is now definitely established as a poison in biological systems. The French Scientist Andre-Marie Ampere suggested around 1811 that anhydrous hydrofluoric acid might be a compound of hydrogen with an unknown element analogous to chlorine for which he suggested the name fluorine. Fluorspar was then recognized as calcium fluoride used as a metallurgical flux since much earlier. The gaseous free element was finally isolated by Henry Moissan in 1886. But upto WWII (1939-45) fluorine remained no more than a lab curiosity with little industrial applications. But since WWII fluorine chemistry was extensively cultivated for war purposes mainly in the separation of uranium isotopes U-238 and U-235. Since then a huge fluorine chemical industry has grown up with fluorinated polymers and fluorinated drugs at the centre stage and a vested interest developed in the US military industrial complex not only to keep much of fluorine chemistry secret but even to distort science in such a way that environmental poisoning can be masqueraded as providing public health services through deliberate drinking water fluoridation and putting fluoride in toothpaste and other dentifices. Fluoride upto about 1940 was universally perceived as a poison. But after WWII it was tainted with a reputation of preventing caries in teeth of children and help children have beautiful smiles. Aluminium extraction from bauxite, nuclear fuel processing and manufacture of phosphate fertilizers all liberate high volumes of toxic fluoride as waste product whose "safe disposal" still now remains an unsolved problem in S&T. But now US Govt. is being forced by conscientious scientists and other sensitive sections of the civil society to stop water fluoridation and its use in toothpastes. A good account can be found in the following recent book (2010):

The Care Against Fluoride

How Hazardous Waste Ended up in our Drinking Water and the Bad Science and Powerful Politics that keep It there Paul Connet, James Beck and H.S. Micklem In India there is no water fluoridation and industrial fluorosis from aluminium extraction, nuclear fuel processing and phosphate fertilizer production as yet are possibly not that serious.

In India more groundwater withdrawal in fluorinated areas, more phosphate fertilizer (usually contain 1 – 3 per cent fluoride) application, more fluoride emitting industries will escalate more environmental fluoride poisoning. So more use of surface water for drinking and more application of biofertilizers will mitigate fluoride poisoning.

Reasons for Slow and Late Development of the Chemistry and Biochemistry of Fluoride

1. Because of chemical peculiarities it is extremely difficult to detect and determine small concentrations of fluoride in solutions and biological samples. Earlier colorimetric methods are neither sensitive nor accurate. The situation have somewhat improved after the discovery of fluoride ion selective electrode only recently.
2. Indian science is not rooted to its soil and its people. It is to be noted that no chemistry book, be it a text-book or comprehensive reference book, contains even mention of the adverse biological effects of fluoride, although much became known since 100 years ago. Medical books contain some discussions but neither they are much, nor they are treated in Indian medical courses and curricula. Reasons are not difficult to guess. Indian academicians frame syllabi, courses and curricula with the help of books prepared in Europe and America for their own students. Arsenicosis and fluorosis there have now become events of distant part. But it is not so in India. Still the subjects do not receive due attention either in Indian chemistry or medical curricula.
3. Some sort of elitism and bias against poor rural people exist in the Indian academic establishment. Adverse health effects due to fluoride and arsenic become more severe for the rural poor, the tribals in particular, because neither they have access to safe water, nor they have adequate nutrition. Proper nutrition to a large extent can countervail the adverse health impacts of fluoride and arsenic. About one third of the capital city of Delhi are constrained to drink high fluoride water. But the situation there though not good but much better than the rural poor of Jharkhand. Good food with proteins, vitamins, minerals, antioxidants countervails the adverse health effects of fluoride.

Chemistry of the Health Hazards of Fluoride Poison

Because of some unique chemical characteristics of fluoride it cannot find any useful role in biochemical machineries of living systems. Hence fluoride was eliminated in the Natural Selection of Chemical Elements in the Origin and evolution of Terrestrial Life which started some 3.6 billion years ago. Still some- not many – scientists in developed countries continue to publish bad data and erroneous evidences purported to show the "essentiality" of fluoride in "some" living systems. Presence of some fluoride in some living systems do not prove their essentiality as presence of some mercury in biological systems do not prove their essentiality.

Fluoride is the Strongest Hydrogen Bonder

Fluoride forms very strong hydrogen bonds with the amide groups of proteins and disrupts their secondary, tertiary, and quaternary structures. Disruptions of the hydrogen bonds cause the secondary and other structures of the proteins to be disrupted which make them fail to perform their proper biological functions. Enzymes are proteins and so enzymes can not catalyse many biochemical reactions and so a host of diseases results. If the shape (or conformation) of the protein is greatly distorted by the fluoride, the body's immune system will no longer be able to recognize the distorted protein as its own and will treat it instead as a foreign protein and attempt to destroy it. The immune response set off by the distorted protein can then be observed as an autoimmune allergic reaction, such as a skin reaction or gastro intentional disturbances.

Much of body proteins (about 30 per cent) are collagen type of proteins. When fluoride disrupts the structures of proteins of the skin wrinkles appear and premature aging sets in. Fluoride also disrupts the hydrogen bonding in DNA of cells and cause genetic damage not excluding cancer.

Fluoride and Hydroxide Ions are Similar in Size and Electronic Structure

Because of this fact F^- and OH^- are exchanged with facility which results in calcification of ligaments and fluorosed bones that cause various arthritic types of syndrome and make the bones fragile. Chemical reactions showing the ion exchanges are as follows :

$Ca_{10}(PO_4)_6(OH)_2$ Calcium hydroxy phosphate or apatite	$\xrightarrow[OH^-]{F^-}$	$Ca_{10}(PO_4)_6(OH)F$ Calcium hdyroxy mono-fluoro-phosphate
	$\xrightarrow[OH^-]{F^-}$	$Ca_{10}(PO_4)_6F_2$ Calcium fluorophosphate or fluorapatite

Since much of the teeth, bones, tendons, cartilages are collagen type of proteins on which calcium hydroxy phosphate, calcium carbonate become slowly and systematically deposited which is called *mineralization* or *calcification*. So fluoride replaces hydroxide ions and the tissues become fluorosed which gives rise to host of diseases called mottled teeth, skeletal fluorosis etc. Arthritis is inflammation of one or more joints, characterized by swelling, warmth, redness of the overlying skin, pain and restriction of movements. Early stages of fluorosis is difficult to distinguish from a host of arthritic diseases. And these are of the greatest concern in fluorosis endemic regions, although these do not manifest in many types of skeletal fluorosis like

Kyphosis (hunchback), Genu valgum (knock-knee), Genu varum (bow-leg) and various deformations of the leg and feet.

Chronic low-level fluorosis cause hips to become fluorosed leading to increased tendency to hip fracture even at lower ages.

Fluoride has a strong tendency to form complexes (or chelates) with many metal ions (like copper, calcium, magnesium, iron, zinc, manganese, chromium) making them less available for proper functioning of many enzymes.

Calcium lost through complex formation cause hypocalcemia, loss of copper through fluoride complex formation lead to malfunctioning of many endocrine glands of which thyroid malfunctioning is very important. Iron complex formation leads to anaemia. And so on.

Fluoride carbon bond is one of the strongest in organic chemistry. For this many modern drugs are systematically fluorinated so as to make them more potent. About 30 per cent of modern drugs are fluorinated. Their adverse effects are yet to be ascertained. Many organic compounds containing carbon-fluorine, phosphorus-fluorine bonds like sarin, tabun are deadly chemical warfare nerve gases.

Fluoride and Defective Child Births

Fluoride affects the brain and leads to neurological disorders and defective childbirths with low IQ.

Birth of children with Down's Syndrome or Mongol child is hastened by fluoride. Mongolism is due to chromosomal aberrations which is normally around 40 as the age of the mother. But it is around 35 when the mother is exposed to 0.1 to 0.5 mg/litre and it is lowered still to around 30 years or less if the fluoride concentration increases to 1.2 to 2.8 mg/litre.

Diverse Other Health Effects

Diverse are the adverse health effects of fluoride in human systems. Not everything is studied as yet. But most important are as has been described earlier are:

- ☆ Immune System Damage
- ☆ Premature Aging
- ☆ Genetic Damage
- ☆ Cancer
- ☆ Increased probability of hip fracture
- ☆ Skeletal Fluorosis
- ☆ Brain Damage and Birth of Children with Down's Syndrome
- ☆ The Thyroid gland, kidneys are affected
- ☆ Male infertility due to deformed sperms

Though the deformed limbs and visible crippling skeletal flurosis are spectacular in sight, but they may be likened to the *tip of the ice-berg*. The sub- clinical non-skeletal fluoroses are much more widespread and more important. An epidemic of arthritis is now sweeping even in USA.

Since there is no possible treatment and recovery impossible through medical intervention, so *prevention is the only option. Fluoride free water drinking and good nutrition can cause some recovery*. For this people must undersand, awareness must be generated. Herein the science people, the medical people in particular, has a social and human responsibility.

Imperative Research Needs

Fluorosis research in India can open up vast fertile fields of research not only for Chemists, Biochemists, Biologists, Medical people, Geologists but also for Sociologists, Economists, Historians and others.

Note: The only useful and readable book from India is the following **A Treatise on Fluorosis** By Prof. (Dr.) A. K. Susheela Revised 3rd Ed. 2007 (price Rs.400/-) E-mail : susheela@bol.net.in

* Co-author of Introduction to Environmental Studies (2002), author of Arsenic in Bengal : Nature and Remediation (2006, 2010) and Fluoride Poisoning : A Medico-Geological Discussion (2011), all in Bengali.

Chapter 6

Physico-Chemical and Bacteriological Investigation of Potable Water of Mizoram

M. Pathak and Hashmat Ali

ABSTRACT

Water pollution has emerged as a major public health concern as a result of man-made and environmental changes at local and global scales. Increasing population, geological factors, rapid urbanisation, agricultural developments, global markets, industrial development and poor wastewater regulation have affected the quantity and the quality of water.

The water samples of different sources from different locations of Lunglei city were collected and analysed for physico-chemical parameters such as pH, turbitity, TDS, electrical conductivity along with Arsenic and Heavy metal ions by following standard methods in the year 2010 during pre monsoon season *i.e.* during February-March. Further, the Arsenic contamination in the potable water of the neighbouring states like Assam and Manipur has been reported off late and it has been found to be maximum at Thoubal district of Manipur(798–986 μg/l) and Jorhat district of Assam (490 μg/l)(1). Thus, we proposed to analyse water samples for Arsenic as well.The results of analysis suggest that the quality of water is well within the safe limits.

Introduction

Lunglei is located in the South-central part of Mizoram in India with a population of 47,355(as per 2001 census).The city being situated in the hilly tracks (average sea-level height 4000ft, Lat- 220 88', Long-92073') of North East India does not have much facility of groundwater exploitation. The potable water used by the inhabitants is usually the rain water which is either collected as run off from the small 'Tuikhurs'(ditches made by local bodies) or is pumped by the public Health Engineering (PHE) department from the catchment ponds made for the purpose.2 The catchment ponds as well as tuikhurs are created at the valleys of hills, so that the rain water is collected after they travel a considerable distance and undergo enough chances of getting exposed to all available substances. The systematic analysis data of portable water of this city is very scanty.

Water pollution is a major global problem. It has been suggested that it is the leading worldwide cause of deaths and diseases(2), and that it accounts for the deaths of more than 14,000 people daily(3). An estimated 700 million Indians have no access to a proper toilet, and 1,000 Indian children die of diarrheal sickness every day(4).

Causes

The specific contaminants leading to pollution in water include a wide spectrum of chemicals, pathogens, and physical or sensory changes such as elevated temperature and occurring (calcium, sodium, iron, manganese, etc.) the concentration is often the key in determining what is a natural component of water, and what is a contaminant. High concentrations of naturally-occurring substances can have negative impacts on aquatic flora and fauna.

Oxygen-depleting substances may be natural materials, such as plant matter (*e.g.* leaves and grass) as well as man-made chemicals. Other natural and anthropogenic substances may cause turbidity (cloudiness) which blocks light and disrupts plant growth, and clogs the gills of some fish species(5). Pathogens - Coliform bacteria are a commonly used bacterial indicator of water pollution, although not an actual cause of disease. Other micro organisms sometimes found in surface waters which have caused human health problems also.

Organic water pollutants include detergents, disinfection by-products found in chemically disinfected drinking water, such as chloroform,Food processing waste, which can include 3 oxygen-demanding substances, fats and grease, Insecticides and herbicides, a huge range of organohalides and other chemical compounds. Arsenic contamination of groundwater is a natural occurring high concentration of arsenic in deeper levels of groundwater, which became a high-profile problem in recent years due to the use of deep tube wells for water supply in the Ganges Delta, causing serious arsenic poisoning to large numbers of people. Arsenic is a carcinogen which causes many cancers including skin, lung, and bladder as well as cardiovascular disease(6).

Although, there is considerable interest among researchers in understanding water pollution and its causes there is surprisingly little work in understanding the linkage between water pollution and health.

Materials and Methods

There is a lot of permutations and combinations possible for water getting polluted with the available factors. In Lunglei,the supply by the PHE department is not sufficient for the inhabitant,so the inhabitants have to resort to other alternatives such as tuikhurs, springs etc. The water samples of bore well (tube well),tuikhurs and PHE supply were collected from different localities and were marked as below:

- ✰ Tube well water sample of Pukpui Veng area –LW1
- ✰ Tuikhur water sample of Serkawn Veng area- LW2
- ✰ Tuikhur water sample of Bazar Veng area- LW3
- ✰ Tuikhur water sample of Rashi Veng area- LW4
- ✰ Tuikhur water sample of Chanmari Veng area- LW5
- ✰ Tuikhur water sample of Farm Veng area- LW6
- ✰ Tuikhur water sample of HranchalKawn Veng area-LW7

The samples were collected in thoroughly washed sample bottles by grab sampling method(7). The pH values of water samples were recorded with the help of digital pH meter-335 while the EC (electrical Conductivity) and TDS (total dissolved Solids) were recorded by using Systronic Conductivity/TDS meter 307.However, the Turbidity of the samples was recorded with the help of Nephelo Turbity Meter -132 using formazine as standard. Coliform is the most common pathogen responsible for water pollution. The most probable number (MPN) of coliforms was determined by multiple tube dilution method. This technique was used to reach first to presumative test which was followed by conformity test.

Presumative Test

Five fermentation tubes were filled with 9ml Mac Konkey broth each for dilution. Dilution of the order 1:1000 was achieved for each. One Durham's vial each was kept in an inverted position in each fermentation tube and each tube was plugged with sterilized cotton. The samples were put inside the fermentation tube with the help of sterilized pipette.The tubes were vigorously shaken and incubated for 48 hours at 35-37°C.The observation of production of gas in Durham;s tube indicated the positive result.All the tubes showed production of gas and were subjected for confirmatory tests.

Confirmatory Test

The fermentative tube was filled with 10 ml of Brilliant Green Lactose Bile (BGLB) broth and Durham's vial was put in an inverted position in each fermentation tube. The tubes with positive 5 results were shaken gently and one loopful of sample was transferred to each fermentative tube having BGLB broth.These tubes were incubated for 48 hours at 35-37°C. The tubes producing gas were treated as positive and MPN per 100 ml was calculated according to the following formula : MPN/100ml = (Table Value of MPN X10)/ starting value.

Arsenic Test

Arsenic was tested in the samples by Field testing method. In this method, Arsenics is tested by reducing inorganic Arsenic to Arsine (ASH_3) with the help of Zinc and Hydrochloric acid. The Arsine gas evolved is made to pass through mercury bromide (HgBr2) indicator paper and the intensity of colour indicated the concentration of Arsenic present in the samples, as shown in the Table 6.2.

Results and Discussion

The pH values varies from 6.4 to 7.2 which are well within the safe limit of 6.5 to 8.5.The data suggest that the pH values are slightly towards alkaline nature which may be due to leaching of ashes produced during preparatory season for Jhum cultivation. The EC

Table 6.1: Physico-Chemical and Bacteriological Parameters of Water Samples

Sample	*Avg. Temp. (0°C)*	*pH*	*EC (milli Sieman/cm*	*TDS (mg/l)*	*Turbidity (NTU)*	*MPN*
LW1	28	6.4	185.5	158	5.7	5
LW2	27	7.2	172.0	115	2.7	28
LW3	26	7.1	190.0	135	1.6	33
LW4	25	7.0	137.0	112	3.0	22
LW5	26	7.1	222.0	180	4.1	24
LW6	27	7.2	168.0	145	2.3	21
LW7	28	7.1	227.0	172	1.2	44

Table 6.2: The Values of Chemical Parameters

Water Samples	*Total Iron (ppm)*	*Total Chloride (ppm)*	*Total Fluoride (ppm)*	*Total Hard-ness*	*Total Arsenic (ppm)*
LW1	28	6.4	185.5	158	5.7
LW2	27	7.2	172.0	115	2.7
LW3	26	7.1	190.0	135	1.6
LW4	25	7.0	137.0	112	3.0
LW5	26	7.1	222.0	180	4.1
LW6	27	7.2	168.0	145	2.3
LW7	28	7.1	227.0	172	1.2

varies from 137 to 227 mS/cm. These values are generally higher for tuikhur samples. This may be attributed to transport phenomenon and subsequent leaching of ions. The TDS values varies from 112 to 180 mg/l which also falls within the desirable limit of 500mg/l. The turbidity values varies between 1.2 and 5.7 NTU. The desirable and permissible limit being 5.6 and 10 NTU respectively.The total coliform density varies from 5 to 44. Compared with BIS (10MPN/100ml), these values are higher in general. This may be attributed to open faecal discharge and existence of Kuchch a/pit latrine, piggeries and poultries(8). The hardness of samples varied from 40 to 120.The concentration of Iron was found to be very low except in LW1.The solubility of Iron mineral increases under reducing conditions enriching the iron concentration in the sample. The Chloride concentration ranges from 10 to 40 ppm which might be arising due to leaching of Sewage water waste and bio-degradation of chlorine containing organic compounds. The fluoride concn is very low which might be due 7 to lack of Fluoride bearing minerals in the Strata through which water is assimilating(9). The Arsenic concentration was found to be very low. Mizoram is having minimum arsenic level among North–Eastern states(10). The minimum level of concn may be due to the presence of thick clay bed which acts as a barrier to prevent the vertical percolation of arsenic.

Conclusion

The findings on physico-chemical and bacteriological studies suggest that the water quality of Lunglei area is free from the impurities and is safe as per BIS as well WHO standard. However, further investigations of other parameters will reinforce the findings.

References

1. Singh, A.K. (2004). Arsenic contamination in groundwater of North Eastern India. In: *National Seminar on Hydrology with Focal Theme on "Water Quality"* held at National Institute of Hydrology, Roorkee during Nov. 22–23.
2. Twarakavi, N.K.C. and Kaluarachchi (2006). Arsenic in the shallow groundwaters of conterminous United States: assessment, health risks, and costs for MCL compliance" J. J. *Journal of American Water Resources Association*, 42(2): 275–294.

3–4. Mukherjee, A., Sengupta, M.K. and Hossain, M.A. (2006). Arsenic contamination in groundwater: A global perspective with emphasis on the Asian scenario. *Journal of Health Population and Nutrition*, 24(2): 142–163.

5. Chatterjee, Amit, Das, Dipankar, Mandal, Badal K., Chowdhury, Tarit Roy, Samanta, Gautam and Chakraborti, Dipankar (1995). Arsenic in groundwater in six districts of West Bengal, India: the biggest arsenic calamity in the world. Part I. Arsenic species in drinking water and urine of the affected people". *Analyst*, 120(3): 643–651.

6. Meliker, Jaymie R. (2007). Arsenic in drinking water and cerebrovascular disease, diabetes mellitus, and kidney disease in Michigan: A standardized mortality ratio analysis. *Environmental Health Magazine*, 2: 4.

7. APHA, AWWA (1998). *Standard Method for Examination of Water and Wastewater*, 20th edn. American Public Health Association, Washington, D.C.

8. Rajurkar, N.S., Nongbri, B. and Patwardhan, A.M. (2003). Physico-chemical and bacteriological investigation of River Umshyrpy at Shillong, Meghalaya. *Ind. J. Environ. Health*, 45: 83–92.

9. Singh, A.K. (2004). Arsenic contamination in groundwater of North Eastern India. In: *Hydrology with Focal Theme on Water Quality*, (Eds.) C.K. Jain, R.C. Trivedi and K.D. Sharma. Allied Publishers, New Delhi, pp. 255–262.

10. Singh, A.K. *et al.* North Eastern Regional Institute of Water and Land Management, Tezpur. *Assesment of Arsenic, Fluoride, Iron, Nitrate and Heavy Metals in Drinking Water of North Eastern India.*

Chapter 7

Effect of Sewage Pollution on the Hydrological Status of Wetlands of Shikaripara, Dumka (Jharkhand)

Hashmat Ali, V.P. Sahay, Santosh Kr. Singh, R.K. Bishen, D.N. Singh and Ashok Kr. Jha

ABSTRACT

The present Chapter throws light on the effects of Sewage Pollution from the drains of surrounding area on the hydrological status of two ecologically different waterbodies at Shikaripara (Dumka) from January 2010 to December 2010. The range of variations for some of the physico-chemical parameters like dissolved oxygen, free CO_2, carbonate, bicarbonate alkalinity, calcium, chloride, phosphate, nitrate and BOD was 3.72-8.42 mg/l, 3.10-5.40 mg/l, 5.95-10.90 mg/l, 102-148 mg/l, 20.90-39.80 mg/l, 17.25-24.40 mg/l, 0.010-0.134 mg/l, 0.220-1.070 mg/l and 1.70-3.60 mg/l in pond-I and 3.10-7.35 mg/l, 3.010-47.55 mg/l, 12.30-21.40 mg/l, 320-460 mg/l, 64.05-100.05 mg/l, 71.50-111.20 mg/l, 0.188-0.850 mg/l, 0.260-1.652 mg/l and 5.20-9.41 mg/l in pond-II respectively. These parameters exhibited a marked difference between two waterbodies depending upon the quantity, quality and nature of sewage pollution.

Introduction

The natural water resources of the country are shrinking fastly due to introduction of inorganic and organic wastes resulting into rapid and progressive eutrophication. According to survey conducted by NEERI, 70 per cent of India's freshwaters are polluted because these water resources have been freely used for the industrial and public wastewater disposal (Basu 1986). All life, from micro-organisms to man, require water but deteriorating quality of water is a serious problem today. Because all water resources have reached to a point of crisis due to unplanned urbanization and mushrooming of industries, it is essential to formulate a sound public policy for water quality improvement. Kant and Vohra (1989) have rightly suggested that the management of any aquatic ecosystem is conservation of freshwater habitat with an aim to maintain the water quality or to rehabilitate the physico-chemical and biological quality of water. It is also necessary to get timely information of pollution status of waterbody before any quality improvement programme of water is planned. However, the need for immediate evaluation and documentation of the water quality of our freshwater resources can hardly be overemphasized in Jharkhand except the work of Bose and Sinha (1981), Sinha (1986), Bose and Lekra (1994), Kumar (1994, 1995, 1996 a). In view of the above, an attempt was made to assess the impact of domestic sewage on certain physico-chemical characteristics of two ecologically different ponds.

Materials and Methods

Monthly sampling of water was carried out in two ponds from January to December of 2010 in order to evaluate the physio-chemical parameters of water. Air and water temperature were recorded by mercury thermometer and pH with the help of pH meter. The transparency was measured by Secchi's disc. Other parameters, such as dissolved oxygen (DO), free CO_2, alkalinity, chloride, phosphate, nitrate, BOD etc, were determined by standard methods (APHA 1995)

Pond-I is situated at Sarasdengal village. This pond is moderately a large shallow waterbody, almost rectangular shaped with an area of 2.5 acres. The water depth of this pond was from 5 to 10 feet. The pond has sand and clay mixed bottom and was having coastal saprophytes. The bank was also provided with some trees. Bathing activities by the local people were usually noticed.

Pond-II is situated near Block office. It was also a large perennial water body, spreading in about 2 acres with an average depth of 7 feet. The adjoining areas of this pond are residential and thickly populated. It is subjected to pollution due to discharge of domestic sewage and by the defaecation of local people. The sewage carrying amount of waters, passing through the residential areas, is discharged into the pond by large drains throughout the period of investigation. The northern bank of this pond is sloppy, covered with variable size of dense green vegetation. It is also infested with macrophytes, especially Eicchornia crassipes.

Results and Discussion

Seasonal fluctuations of certain abiotic factors of pond waters have been shown in Tables 7.1 and 7.2. During the present investigation, the minimum temperature was recorded in January (16.5°C in pond-I and 16.2°C in pond-II). Whereas its maximum temperature was observed in June (32.5°C in pond-I and 34.0°C in pond-II) of the study period. Transparency was influenced mainly by suspended organic matter (Green 1974). A higher value of transparency was recorded in pond-I (31.3-47.5 cm) than the pond-II (29.2-37.7 cm). It may be due to minimum input from the surrounding. The low transparency in pond-II might be due to various human activities and domestic effluents from the adjoining areas. In the present study, higher transparency was recorded in winter months in both the ponds. This could be attributed to less decomposition of organic matter due to low temperature and less input of solids by the surface run off due to cessation of monsoon rain. This confirms the findings of George (1976). But Kaushik *et al.* (1991) observed higher transparency during summer in Matsya Sarovar at Gwalior. Low transparency was recorded during later summer and monsoon which may be due to influx of rain water in addition to huge suspended colloidal matters. Kumar *et al.* (1996) observed that domestic sewage is also responsible for low transparency.

pH is the index of water quality. Free CO_2 and total alkalinity show direct impact upon the status of pH. During the present investigation, pH value ranged between 7.5 and 8.4 and 8.4 and 7.9 and 8.4 in pond-I and II respectively. The pH values showed that the both ponds were alkaline in nature.

Low pH was observed during summer months which might be due to the surplus amount of free CO_2 on account of accelerated rate of decomposition during periods of high temperature. pH value in

Table 7.1: Monthly Variations in Physico-chemical Characteristics of Two Ecologically Different Ponds at Shikaripara (Dumka) Jharkhand from January to December 2010

Month	*Pond*	*Water Temp (°C)*	*Trans-parency (cm)*	*pH*	*DO*	*Free CO_2*	*Carbo-nate*
January	I	16.6	42.5	8.0	7.00	-	8.00
	II	16.3	30.8	8.2	4.20	25.40	
February	I	22.4	41.8	8.1	6.80	-	10.80
	II	20.2	30.6	8.1	4.52	22.40	
March	I	27.2	41.2	8.3	6.05	-	11.60-
	II	26.7	30.2	8.3	7.35	48.50	
April	I	28.9	39.0	7.2	5.20	-	6.75
	II	28.2	30.2	8.0	4.05	-	18.50
May	I	31.3	38.0	7.6	5.12	-	7.00
	II	30.0	29.8	8.0	4.50	-	15.50
June	I	32.8	38.5	7.7	4.50	4.55	
	II	34.0	30.0	7.9	3.25	-	12.65
July	I	32.3	31.3	7.5	3.75	5.90	-
	II	31.7	29.3	7.2	3.20	7.50	-
August	I	31.4	33.5	7.8	4.15	4.00	-
	II	31.2	32.1	8.0	3.15	18.50	-
September	I	30.2	36.0	8.0	5.35	5.35	-
	II	30.3	34.2	8.1	4.20	3.20	-
October	I	29.5	44.3	8.2	6.80	3.25	-
	II	28.8	34.1	8.2	5.35	-	-16.20
November	I	23.3	45.4	8.2	7.45	-	9.20
	II	23.1	36.2	8.3	4.70	-	14.50
December	I	18.6	46.6	8.1	8.40	-	10.30
	II	18.8	38.2	7.8	4.95	-	21.85

All parameters values are in mg/l except Temp. & pH.

pond-II is comparatively higher than the pond-I which could be attributed to the rich growth of aquatic plants than remove free CO_2 from the water during day time and high value of carbonates and bicarbonates in the water.

Table 7.2: Monthly Variations in physico-chemical Characteristics of Two Ecologically Different Ponds at Shikaripara (Dumka) Jharkhand from January to December 2010

Month	*Pond*	*Bicarbo-nate*	*Calcium*	*Chloride*	*Phos-phate*	*Nitrate*	*BOD*
January	I	130	31.60	20.65	0.035	0.234	2.10
	II	330	90.35	72.52	0.200	0.545	5.25
February	I	150	39.80	21.22	0.032	0.325	2.20
	II	352	91.35	69.22	0.418	0.302	6.40
March	I	138	32.55	22.75	0.065	0.633	2.95
	II	410	95.30	88.95	0.602	0.860	8.35
April	I	148	31.10	21.15	0.055	0.572	2.90
	II	428	96.60	100.15	0.855	1.645	8.60
May	I	145	26.35	17.50	0.130	0.460	2.90
	II	468	100.10	112.25	0.345	1.670	9.55
June	I	135	32.50	20.20	0.138	0.730	2.90
	II	446	93.90	87.15	0.720	1.450	8.20
July	I	108	23.15	20.75	0.045	0.805	3.40
	II	370	77.35	78.15	0.502	0.833	6.70
August	I	125	23.15	22.50	0.012	1.082	3.50
	II	312	70.75	76.45	0.512	0.910	7.80
September	I	114	22.40	23.60	0.012	0.855	3.70
	II	298	64.20	73.35	0.495	0.670	5.30
October	I	110	21.75	23.65	0.020	0.455	3.15
	II	338	73.45	71.35	0.435	0.612	6.25
November	I	110	24.60	24.95	0.028	0.282	2.20
	II	366	77.20	76.30	0.322	0.442	6.35
December	I	125	28.10	21.40	0.033	0.220	1.90
	II	840	90.60	74.15	0.215	0.560	5.20

All parameters values are in mg/l.

The dissolved oxygen ranged between 3.75 and 8.40 mg/l pond-I and between 3.20 to 7.35 mg/l in pond-II. The levels of DO in a waterbody are perhaps of the greatest importance for the survival of the aquatic organisms. During the present investigation, both seasonal as well as spatial changes in oxygen content have been recorded. The general trend of changes in DO concentration during different seasons is directly or indirectly governed by fluctuations of temperature and BOD. Higher values of DO content was recoded in winter (pond-I), the period during which the water temperature was lowest. This might be due to the fact that the solubility of DO increases with the decrease in water temperature. Kumar (1996 b) has reported low solubility of oxygen at higher temperature, which is also in close accordance with the present study. Loss of oxygen to the atmosphere and its utilization by faster decomposition of organic matter at higher temperature seems to be the cause for such as observation. Low value of DO in the same pond during monsoon is attributed to the reduction in photosynthetic activity and decomposition of organic matter. DO content was noticed comparatively low in pond-II throughout the study period which might be due to the high rate of oxygen consumption by oxidisable matter coming in along along with the domestic sewage. This is further supported by Singh (1995) and Kumar (1996 b).

Free CO_2 was not recorded throughout the period of investigation in pond-I, it was present from June to October and the range was from 3.25 to 5.90 mg/l. Its presence during monsoon months may be due to the rain water and respiratory activities of aquatic biota. Pond-II was infested with macrophytes, therefore, the presence of free carbon dioxide in water has fluctuated irregularly from 3.20 to 48.50 mg/l. The higher value in pond-II might be due to decomposition of algae and domestic sewage coming along with nearby drains. The carbonate was not always detected during the study period. The range of carbonate varied from 6.75 to 10.58 mg/l and from 12.65 to 21.85 mg/l in pond-I and II respectively. The bicarbonate value was recorded between 108 and 150 mg/l in pond-I whereas its range was between 298 and 440 mg/l in pond-II. The presence of higher total alkalinity in pond-II might be due to rich nutrients with higher productivity (Kumar *et al.*, 1996).

The value of calcium ranged between 21.75 and 32.85 mg/l and 64.20 and 100.10 mg/l in pond-I and II respectively. The concentration of calcium was recorded higher in pond-II which may be attributed to the heavy discharge of organic matter through domestic sewage whereas pond-I received comparatively very little domestic sewage. The gradual rise in the calcium contents from

January to May in pond-II might be due to the rapid oxidation of organic matter in the substrate.

Contamination of water from domestic sewage can be monitored by the chloride assays for the concerned waterbodies. This is because human and animal excretions contain an average of 5 g chloride/Titre (Singh 1997). During the present study, minimum content of chloride was recorded in May (17.50 mg/l) and January(72.56 mg/l) while maximum was recorded in the month of November (24.90 mg/l) and May (112.25 mg/l) in pond-I and II respectively. The maximum value of chloride content in pond-II can be directly correlated with high degree of sewage discharge and human interferences. This is further strengthened by its lower concentration in pond-I.

Thresh *et al.* (1944) advocated that the presence of high chloride is the index of pollution of animal origin. Klein (1957) was also of the opinion that the chlorides appear in aquatic environment mainly due to the sewage contamination. The higher concentration of chloride during summer (May) in pond-II may be associated with reduced water level and frequently run-off loaded with contaminated water from the surrounding settlement and high rates of evaporation. Sunder (1988) and Kumar (1995) have also observed the same pattern.

The phosphate content is considered to be nutrient of major importance in production process (Vollenwider 1968). But Jones and Lee (1982) have conveyed that an emphasis on phosphorus control will have to be given while evaluating eutrophication control options because of its ability of stimulate algal growth as well as the growth of other aquatic plants. The increased application of fertilizers, use of detergents and domestic sewage play a great role in contributing the heavy loading of phosphorus in water (Golterman 1975).

During the present investigation, the phosphate contents was found comparatively higher (0.198-0.855 mg/l) in pond-II, because of this waterbody receives the influx of sewage effluents and decomposed organic matter. The values of phosphate were also observed fluctuating from season to season. Its values were maximum during summer months. The increased solar radiation might have encourage the biological degradation of organic matter and subsequent release of more phosphate.

The nitrate content is an excellent parameter to judge the organic pollution and it represents the highest oxidized form of nitrogen. The range of nitrate value was recorded between 0.250 and 1.088 mg/l in pond-I and between 0.265 and 1.670 mg/l in pond-II. During the present study the higher value of nitrate was recorded during rainy season which might be due to influx of nitrogen rich flood water that brings large amount of contaminated sewage water. However, the nitrate values were sufficiently found higher in summer season in pond-II which might be due to the increased rate of decomposition of organic matters in account of high temperature and concentration of sewage discharge.

During the present investigation, the range of BOD was from 1.90 to 3.40 mg/l in pond-I and from 5.20 to 9.55 mg/l in pond-II. Higher value of BOD was observed during monsoon in pond-I because of input of organic wastes and enhanced bacterial activity, and heavy value of BOD during summer could be attributed to high bacterial activity and heavy input of organic matter. Its lower value during winter might be due to retarded bacterial c=activity affected by decreased light intensity and temperature. Similar observations are also made by Paramasivam and Sreenivasan (1981) Sinha *et al.* (1990) and Kumar (1997).

The overall picture emerged during the present investigation shows that the sewage pollution has become a great problem in Jharkhand. Thus, the present study may appear as simple one, but the results are the indicative of a big social issue, specially in that area which is poor economically but trubulent politically. Such small ecological problems may add fuel to the fire that can ultimately upset the whole nation. Thus, the crying need of time is the proper environmental management and mass awakening among the people right from the grass root level to the top ecoplanners. Only then our long cherished dream of establishing the "Ecological Socialism" on the neglected land of Santal Parganas of Jharkhand could come true.

Acknowledgement

The author is grateful to Prof. P. K. Ghosh, Head, P. G. Deptt. of Chemistry, S. K. M. University, Dumka for encouragements and to Dr. S. P. Yadav, Principal, S. P. College, Dumka for providing Laboratory facilities.

References

1. APHA (1985). *Standard Methods for the Examination of Water and Wastewater*. American Public Health Association Inc., New York, p. 268.
2. Basu, B. (1986) Environment protection – a many – faceted problem. *Yojun*, 30: 4–6.
3. Bose, S.K. and Lakra, M.P. (1994). Studies on the macro-zoobenthos of two freshwater pinds of Ranchi. *J. Freshwat. Bio.*, 6: 135–142.
4. Kant, S. and Vohra, S. (1990). Lakes their management and conservation. In: *Management of Aquatic Ecosystem*, (Ed.) V.P. Agarwal, B.N. Desai and S.A.H. Abidi. Society of Bioscience, Muzaffarnager, p. 155–168.
5. Kumar A 1996 a. The limnological profiles of a tropical fish farming pond at Dumka (Santal Pargana), Bihar. 3. Ecobiol. 8.: 117–122.
6. Kumar, A. (1996b). Impact of industrial pollution on the population status of the endangered Gangetic dolphin (Platanista gengetic) in the river Ganga, Bihar, India. *Poll. Arch. Hydrobiol.*, 43: 469–476.
7. Mukherjee, A., Sengupta, M.K. and Hossain, M.A. (2006). Arsenic contanuation in groundwater. *Journal of Health Population and Nutrition*, 24(2): 142–163.
8. Singh, M. (1995). Impact of human activities on the physico-chemical conditions of two fish ponds at Patna, Bihar. *J. Freshwat. Biol.*, 7: 13–17.
9. Singh, A.K. (2004). Arsenic contamination in groundwater of North Eastern India.
10. Singh, M.P., Sinha, K., Sinha, R. and Mehrotra, P.N. (1994). Limnobiotic investigation of a tropical freshwater reservoir. *J. Freshwat. Biol.*, 6: 215–219.

Chapter 8

Arsenic Contamination in Groundwater of North India

K.M. Singh and Bhartendu Prasad

ABSTRACT

Groundwater arsenic contamination and sufferings of people have been reported in 20 countries from different parts of the world. The magnitude is considered highest in five Asian countries and the severity is in order of Bangladesh> India> Mangolia> China> Taiwan. In all these countries, more and more groundwater withdrawal is taking place because of increase in agricultural irrigation. In India after West Bengal and the bordering districts of Bangladesh, arsenic in groundwater was detected in part of Assam, Arunachal Pradesh, Manipur, Nagaland and Tripura.

Introduction

Arsenic (As) is introduced into soil and groundwater during weathering of rocks and minerals followed by subsequent leaching and runoff. It can also be introduced into soil and groundwater from anthropogenic sources. Many factors control arsenic concentration and transport in groundwater, which include: Red-ox potential (Eh), adsorption/desorption, precipitation/dissolution, Arsenic speciation, pH, presence and concentration of competing ions, biological transformation, etc.

In India, since the groundwater arsenic contamination was first surfaced from West-Bengal in 1983, a number of other States, namely;

Jharkhand, Bihar, Uttar Pradesh in flood plain of the Ganga River; Assam and Manipur in flood plain of the Brahamaputra and Imphal rivers, and Rajnandgaon village in Chhattisgarh state have chronically been exposed to drinking arsenic contaminated hand tube-wells water above permissible limit of 50 μg/L. Many more North Hill States in the flood plains are also suspected to have the possibility of arsenic in groundwater.

Arsenic groundwater contamination has far-reaching consequences including its ingestion through food chain, which are in the form of social disorders, health hazards and socioeconomic dissolution besides its sprawling with movement, and exploitation of groundwater. The food crops grown using arsenic contaminated water are sold off to other places, including uncontaminated regions where the inhabitants may consume arsenic from the contaminated food. This may give rise to a new danger.

The present Chapter thus deals mainly with: (*i*) up to date status of arsenic menace in India, (*ii*) preventive and corrective measures taken so far and results thereof, (*iii*) shortcomings, and possibility of employing success stories of one place to another region, (*iv*) further work to be undertaken.

Arsenic Menace in India

Since groundwater arsenic contamination was first reported in year 1983 from 33 affected villages in four districts in West-Bengal, the number of villages has increased to 3417 in 111 blocks in nine districts till 2008 in West Bengal alone. In 1999, the arsenic groundwater contamination and its health effects in Rajnandgaon district of Chattisgarh state were also detected. In 2002, two villages, Barisban and Semaria Ojhapatti, in Bhojpur district, located in the western part of the Bihar state, were reported having contamination exceeding 50μg/L. As of 2008, out of 38 districts of Bihar, 57 blocks from 15 districts having total population nearly 10 million have been reported affected by arsenic groundwater contamination above 50μg/L. During 2003, 25 arsenic affected villages of Ballia district in Uttar Pradesh and people suffering from skin lesions came into limelight. During 2003-2004, the groundwater arsenic contamination and consequent suffering of hundreds of people were reported from 17 villages of the Sahibgunj district of Jharkhand state, in the middle Ganga plain. In 2004, arsenic concentration was also reported from

Assam in pockets of 2 districts. In 2007, Manipur state, one of the seven North Hill States, had also come into limelight of arsenic contamination in groundwater. Many more North Hill States in the flood plains are also suspected to have the possibility of arsenic in groundwater. As of 2008, West Bengal, Jharkhand, Bihar, Uttar Pradesh in flood plain of Ganga River.

Bihar

Groundwater arsenic contamination in Bihar first surfaced in the year 2002 from two villages, Barisbhan and Semaria Ojhapatti in the Bhojpur district located in the flood-prone belt of Sone-Ganga. A number of scientific studies, focusing mainly on physicochemical analyses of arsenic contaminated groundwater, assessment of extent, mobilization pathways, and possibility of tapping deeper aquifers, arsenic in food chains and its effect on health, were initiated by state and Central government organizations and by different academic institutions working in the State. In addition to R and D studies and exhaustive investigations, Govt. of Bihar, has started a number of schemes, as the precautionary measures to ensure supply of risk-free potable groundwater particularly, in community based localities, and as counteractive steps to combat probable arsenic related threats. As an outcome of scientific investigations and surveys, by 2010, out of 38 districts in the state, 15 districts covering 57 blocks, have been identified as groundwater arsenic contamination above 50 µg/L.

In Bihar, the general awareness of populace about groundwater arsenic contamination and its effects is very less. And people are unaware that the skin skeletal and other health related diseases, experienced by them, are of water origin. Rural people have some phobia of not switching over from habitual use of groundwater to alternate surface sources of water. Therefore, there is a need of breaking such orthodox approach by mass awareness programmes. The corrective and precautionary measures, initiated by the Govt. of Bihar, are too less to the scale up of the problem. and to understand the problem resolving issues, counteractive measures, etc., in comparison to the State of West Bengal. While characteristics and features of the problem, geological formations and causes of the problem are largely similar and represent the hydro-geological setups of the same river basin, except the difference in socio-economic, socio-cultural and social composite structure.

Summary and Conclusions

In India, seven states namely, West-Bengal, Jharkhand, Bihar, Uttar Pradesh in the flood plain of Ganga River; Assam and Manipur in the flood plain of Brahamaputra and Imphal rivers and Rajnandgaon village in Chhattisgarh state have so far been reported affected by arsenic contamination in groundwater above the permissible limit of 50 μg/L. People in these affected states have chronically been exposed to arsenic drinking arsenic contaminated hand tube-wells water. With every new survey, more arsenic affected villages and people suffering from arsenic related diseases are being reported, and the problem resolving issues are getting complicated by a number of unknown factors. It is now generally accepted that the source is of geological origin and percolation of fertilizer residues may have played a modifying role in its further exaggeration. Identification of parental rocks or outcrops is yet to be recognized, including their sources, routes, transport, speciation and occurrence in Holocene aquifers along fluvial tracks of the Ganga-Brahmaputra-Barrak valley. It is reported that the contaminated waters are enriched in Fe, Mn, Ca, Mg, bicarbonates, and depleted in sulphate, fluoride, chloride; pH ranged from 6.5 to 8; redox condition usually in reducing; high on organic matter content; lodged mostly in sand coatings, or sorbed on clays, HFOs, and organic matters. It has been proved that arsenic has affinity with iron in groundwater both positively and negatively, depending upon the condition. This gives a positive hope of devising in situ remediation of the problem of As contamination by removal of Fe from groundwater before withdrawal. Varieties of arsenic removal devices have been developed, based on different working principles, and have been extended to fields. Many of those could not produce satisfactory performance or failed due to lacks in O and M or due to sludge disposal problems. Among the various removal technologies, lime softening and iron co-precipitation have been reported to be the most effective removal technologies, and observed running satisfactorily, where operation and maintenance problems were taken care of by public-private partnership. The available arsenic removal technologies require refinement to make them suitable and sustainable for their large scale effective uses.

Surface waters are free from arsenic contamination. The usages of surface water sources with minor treatment through organized piped water supply system has been proven to be a feasible solution to supply potable water in many places in West Bengal, where surface water availability is assured. Deeper aquifers underneath the contaminated shallow aquifers are found free from arsenic contamination. The deeper aquifers, which are risk free from future threat of contamination from the overlain aquifer, can provide a sustainable source of potential groundwater withdrawal. Involvement of the society in the O and M and making society responsible and knowledgeable can solve many problems associated with the water scarcity issues in the arsenic affected areas.

References

1. Acharyya, S. K., Lahiri, S., Raymahashay, B. C. and Bhowmik, A. (1993). Arsenic toxicity of groundwater in parts of the Bengal basin in India and Bangladesh: The role of quaternary stratigraphy and holocene sea-level fluctuation. *Environ. Geol.,* 39: 1127–1137.

2. Acharyya, S.K., Chakraborty, P., Lahiri, S., Raymahashay, B.C., Guha, S. and Bhowmik, A. (1999), Arsenic poisoning in the Ganges delta. *Nature,* 401: 545–546.

3. Amaya, A. (2002). Arsenic in groundwater of alluvial aquifers in Nawalparasi and Kathmandu districts of Nepal. Master's thesis, Dept. of Land and Water Resources Engineering, Kungl Tekniska Hogskolan, Stockholm.

4. Basu, B.B. (2003). Lessons from arsenic technology Park and critical issues of arsenic mitigation. School of Fundamental Research, Kolkata.

5. Bhattacharyya, R., Jana, J., Nath, B., Sahu, S.J., Chatterjee, D. and Jacks, G. (2005). Groundwater As mobilization in the Bengal Delta Plain. The use of ferralite as a possible remedial measure: a case study. *Appl. Geochem.,* 18: 1435–1451.

6. Bose, P. and Sharma, A. (2002). Role of iron in controlling speciation and mobilization of arsenic in subsurface environment. *Water Research,* 36: 4916–4926.

7. Chakraborti, D. *et al.* (2003). Arsenic groundwater contamination in Middle Ganga Plain, Bihar, India: A future danger? *Environ. Health Perspect.,* 111: 1194–1201.

8. Majumdar, P.K., Ghosh, N.C., and Chakravorty, B. (2002). Analysis of arsenic-contaminated groundwater domain in the Nadia district of West Bengal (India), Special issue: Towards Integrated Water Resources Management for Sustainable Development. *Hydrological Sciences– Journal*, 47(S): S55–S66.

9. Mandal, B.K., Chowdhury, T.R., Samanta, G., Mukherjee, D.P., Chanda, C.R., Saha, K.C., Chakraborti, D. (1998). Impact of safe water for drinking and cooking on five arsenic-affected families for 2 years in West Bengal, India. *The Science of Total Environment*, 218: 185–201.

Chapter 9

Arsenic Contamination of Groundwater and Health Problem in Gangetic Plain of Bhojpur District (Bihar)

Saiyad Rafat Imam and K.M. Singh

ABSTRACT

Arsenic in Water and its physiological impacts is of serious concern worldwide. The number of people at risk of arsenic poisoning through groundwater may be considerably larger than previously thought of detection of arsenic in water sample containing 1861ppb in Pandey tola village in Barhahra block of Bhojpur district, a situation far more serious than represented by the much-touted village Simaria-ojhapatti in Shahpur block of this district has made the matter serious. Analyses of 6317 tube well water samples revealed that arsenic concentration in 38.08 per cent exceeded 10 µg/L, in 19.87 per cent, >50 µg/L and in 4.56 per cent, >300 µg/L limits. Arsenic concentrations up to 3192 ppb were observed. A detailed study in three administrative units within Bhojpur district, i.e block, Gram Panchayet, and village was carried out to assess the magnitude of the contamination. Before our survey most of the affected villagers were not aware that they were suffering from arsenical toxicity through contaminated drinking water. A Preliminary clinical examination in 8 affected villages revealed typical arsenical skin lesions ranging from melanosis, keratosis to Bowens (Suspected). Out of 1322 arsenic patients screened, no. of registered patients with clinical manifestations were 161.

Out of 603 samples of hair, nail and urine analyzed, the per cent age of arsenic concentration above normal level in hair 82 per cent, in nail 89 per cent and in urine 91 per cent detected. The urine, hair and nail concentrations of arsenic correlated significantly (r = 0.72-0.77) with drinking water arsenic concentrations up to 3192ppb. The similarity to previous studies on arsenic contamination in West Bengal and Bangladesh indicates that people from a significant part of the surveyed areas in Bihar are suffering and this will spread unless drives to raise awareness of arsenic toxicity are undertaken and an arsenic safe water supply is immediately introduced.

Introduction

In 1976 arsenic contamination was reported in Punjab, Haryana, Himachal Pradesh and Utter Pradesh, Northern India (Datta and Kaul, 1976). In 1984 groundwater arsenic contamination was discovered in lower Ganga plain of West Bengal (Garai *et al.*, 1984). Only in 1995 was the arsenic situation in West Bengal and consequent suffering of people brought to light (Chakraborti *et al.*, 2002). In 1992 arsenic groundwater contamination in Bangladesh was reported (Dhar *et al.*, 1997). In 2001 groundwater arsenic contamination in the lower plain area (Terai) of Nepal came to notice (Shrestha *et al.*, 2003). In June 2002, arsenic contamination in Bihar in middle Ganga plain and at the same time apprehended contamination in Uttar Pradesh lying in middle and upper Ganga plain was detected (Chakraborti *et al.*, 2003). In January 2004, groundwater arsenic contamination in 25 per cent of 698 hand tube wells from 17 villages of the Sahibgunj district of Jharkhand state, India, in the middle Ganga plain came to notice (Chakraborti *et al.*, 2004). In 2005, Bhattacharjee *et al.* (2005) reported arsenic contamination in three out of nine blocks surveyed in Sahibgunj district, Jharkhand. In February 2004, 26 per cent of 137 hand tube wells were analyzed in 2 district in Assam and had an arsenic contamination above 50 µg/L (Chakraborti *et al.*, 2004). According to an estimation, parts of all the states and countries surveyed in the Ganga-Meghna-Brahmaputra (GMB) plain, which has an area of approximately 500,000 km^2 and a population over 500 million, are at risk from groundwater arsenic contamination (Chakraborti *et al.*, 2004). During March 2005- August 2007, we surveyed Bhojpur district of Bihar. The principal objective of this Chapter is to emphasize the gravity of the present arsenic contamination situation in Bhojpur district based on arsenic analysis of hand tube well water and biological samples, clinical examination, and to make people aware stressing the need for immediate remedial action to avert further worsening of the situation.

Findings

Preliminary Groundwater Arsenic Contamination Study in Bhojpur District of Bihar

Based on 6317 water sample analysis from 122 villages of 1 district, Figure 9.1(D) shows the distribution of tube wells with different arsenic concentration ranges. Of the total, 38.08 per cent had arsenic above 10 µg/L (WHO Provisional guideline value), 19.87 above (Indian standard for arsenic in drinking water) and 4.56 per cent above 300 µg/L of arsenic.

Arsenic Contamination in Bhojpur District

Barhahra consists of 10 villages. Sample from 341 hand tube wells (about 80 per cent of the total number) covering all the villages were analyzed. The distribution [Figure 9.1(A)] indicates that, in Barhahra, 91.8 per cent had arsenic above 10 µg/L, 16.1 per cent between 10 and 50 µg/L, 75.4 per cent above 50 µg/L and 41 per cent above 300 µg/L.

Arsenic Contamination in Panditpur Village, One of the Affected Villages in Bhojpur

All the existing 55 tube well in the village were analyzed for arsenic. The distribution of arsenic concentration indicates that none of the tube wells was safe according to the WHO guideline value, with 3.6 between 10 and 50 µg/L, 96.4 per cent above 50 µg/L and 81.8 per cent above 300 µg/L. This shows that Pandeytola village is highly arsenic contaminated.

Results and Conclusions

In Bangladesh and West Bengal at present, fever people are drinking arsenic-contaminated water than before due to growing awareness and access to arsenic safe water. But in Bihar, Utter Pradesh, Jharkhand, and Assam, villagers are still drinking contaminated water as this problem is largely unrecognized. Since most of these states remain unexplored with respect to arsenic

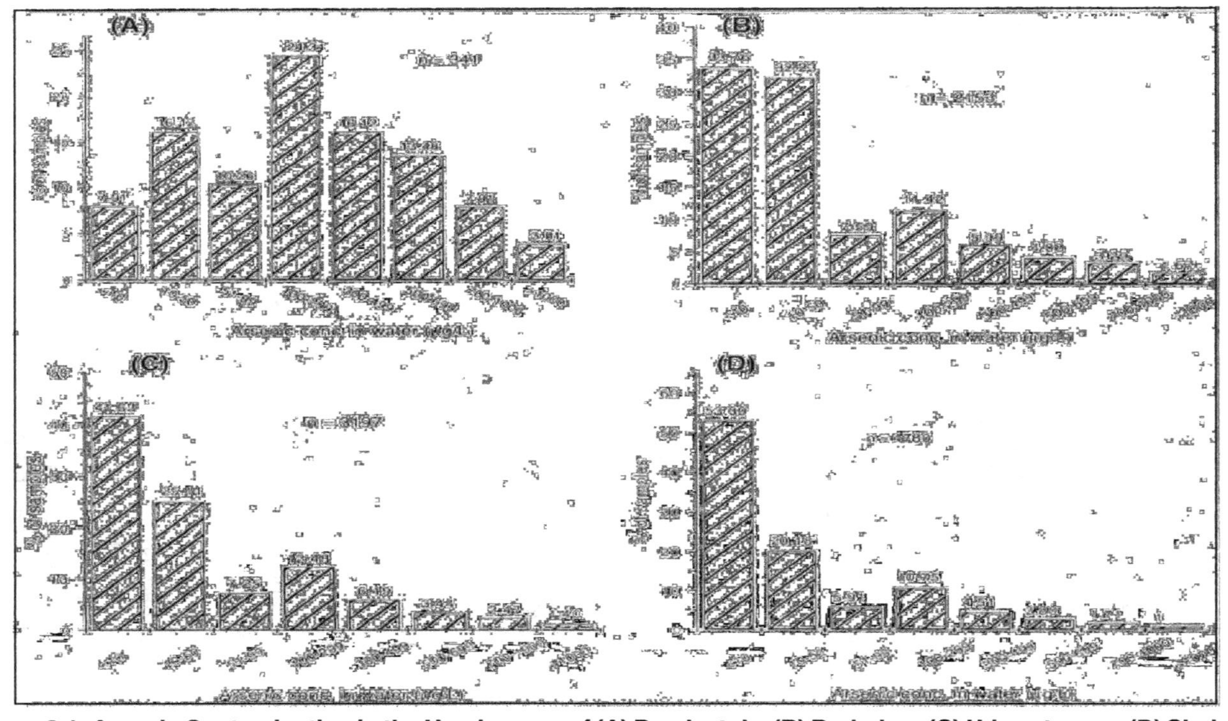

Figure 9.1: Arsenic Contamination in the Handpumps of (A) Pandeytola, (B) Barhahra, (C) Udwantnagar, (D) Shahpur

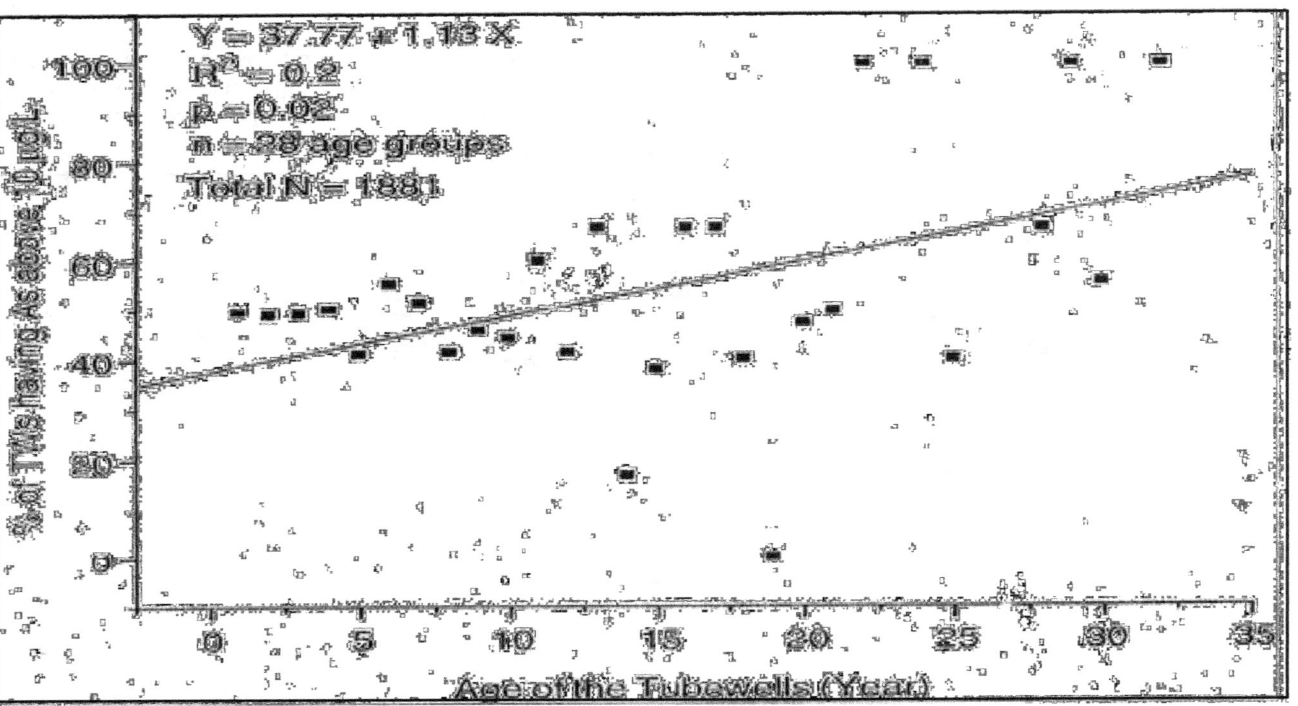

Figure 9.2: Relationship Between Age of the Tubewells and Arsenic Concentrations in Surveyed Areas of Bihar

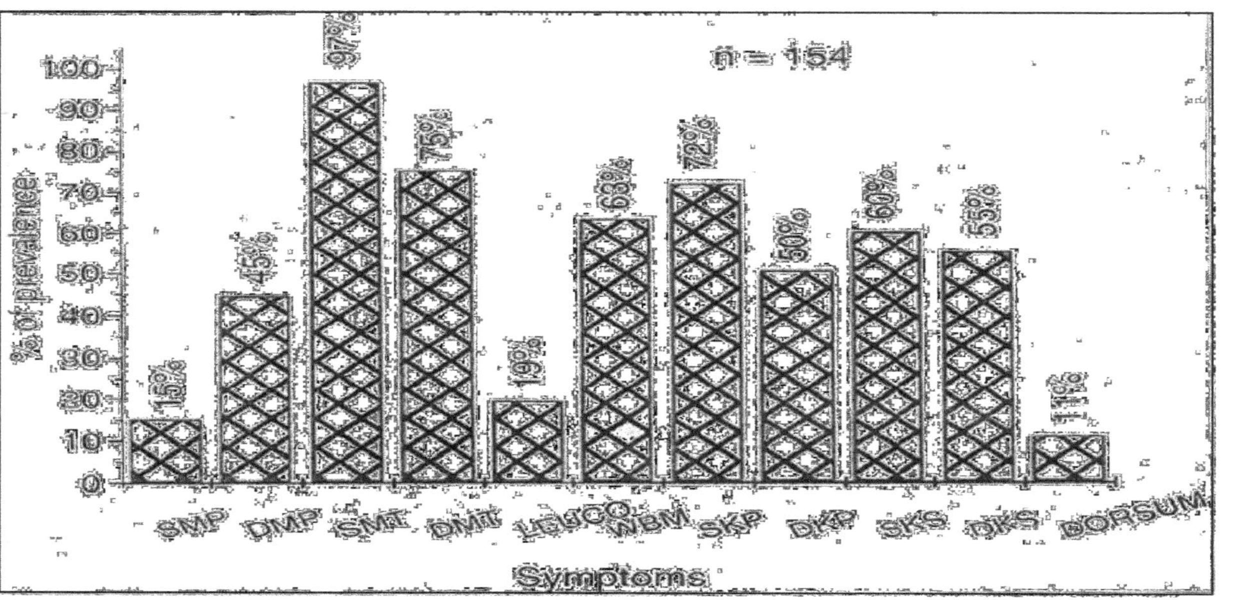

Figure 9.3: The Prevalence of Skin Lesions Observed in 11 Villages of Bhojpur

SMP: Spotted melanosis on plan; DMP: Diffuse melanosis on palm; SMT: Spotted melanosis on trunk; DMT: Diffuse melanosis on trunk; LEUCO: Leuco-melanosis; WBM: Whole body melanosis SKP: Spotted keratosis on palm; SKS: Spotted keratosis on sole; DKS: Diffuse keratosis on sole; DORSUM: Dorsal keratosis; n: Number if arsenic affected patient.

Figure 9.4: Three Patients (A) with Several Arsenical Skin Lesion on Palm and Sole; (B) Arsenical Skin Lesions Along with Multiple Bowens (Suspected); (C) Arsenical Skin Lesions Along with Bowens (Suspected), and Cancer (Suspected), Respectively

contamination, a more intensive survey on a much larger scale is required to ascertain the extent of arsenic affected regions. Early screening in will expedite remedial measures and thus mitigate suffering. The mitigation strategy needs to be location specific, depending on the availability of arsenic-safe option. The tube wells drawing from deeper aquifers free from arsenic and other water-borne contaminations can be used on a community basis. It is well established that, in the Gangetic plain, arsenic contamination in hand tube wells has been observed to decrease below a certain depth (Roychowdhury *et al.*, 1999) but, in unconfined aquifers,even if the tube well is constructed properly. The safety of deep tube wells is dependent on subsurface geology. We have reported some arsenic contamination (Chakraborti *et al.*, 1999) in even deep (depth range 100-415 m) tube wells in Bangladesh, as did also the BGS (BGS/ DPHE, 2001). But these tube wells need to be tested on a regular basis since there are chances of temporal safe water option such as surface water, dug-wells and rainwater harvesting may also be explored, with measure against bacterial and about the arsenic problem and adequate supply of arsenic-safe water to the affected population is required.

References

1. Abernathy, C.O., Marcu, W., Chen, C., Gibb, H. and Write, P. (1998). Report on arsenic work group meeting. Office of Drinking Water, Office of Research and Development,USEPA, Memorandum to Cork, P., Preuss, P. Office of Regulatory Support and Scientific Management, USEPA.

2. Ahmad, S., Sayed, M.H., Barua, S., Khan, M.H., Faruquee, M.H. and Jalil, A. *et al.* (2001). Arsenic in drinking water and pregnancy outcome. *Environ Health Perspect*, 109: 629–631.

3. Arnold, H.L., Odam, R.B. and James, W.D. (1990). Disease of the skin. *Clinical Dermatology*, 8th Edn. Philadelphiap: WB. Saunders, p. 121–2. BAMWSP website (http/ www.bamwsp.org). Last accessed on 17/05/06.

4. Bhattacharjee, S., Chakravarty, S., Maity, S., Dureja, V. and Gupta, K.K. (2005). Metal contents in the groundwater of Sahebgunj district, Jharkhand, India, with special reference to arsenic. Chemosphere 2005, 58: 1203–17. BGS/DPHE. In arsenic contamination of groundwater in Bangladesh. In: Kinniburgh DG, Smedley PL, editors. Final report, BGS technical report WC/ 00/19. Keyowrth, U.K.: Brittish Geological Survey, 2001.

5. Biswas, B.K., Dhar, R.K., Samanta, G., Mandal, B.K., Faruk, I., and Islam, K.S. *et al.* (1998). Detailed study report of Samta one of the arsenic affected villages of Jessore district, Bangladesh. *J. Curr. Sci.*, 74(2): 134–145.

6. Chakraborti, D., Biswas, B.K., Basu, G.K., Chowdhury, U.K., Roy Chowdhury, T. and Lodh, D., *et al.* (1999). Possible arsenic contamination free groundwater source in Bangladesh. *J. Surf. Sci. Technol.*, 15: 180–188.

7. Chakraborti, D., Basu, G.K., Biswas, B.K., Chowdhury, U.K., Rahman, M.M. and Paul, K., *et al.* (2001). Characterization of arsenic bearing sediments in Gangetic delta of West Bengal, India. In: *Arsenic Exposure and Health Effects,* (Eds.) W.R. Chappell, C.O. Abernathy and R.L. Calderon, Elsevier Science, Amsterdam, p. 27–52.

8. Chakraborti, D., Rahman, M.M., Paul, K., Chowdhury, U.K., Sengupta, M.K. and Loth, D. *et al.* (2002). Arsenic calamity in the Indian sub-continent what lesions have been learned? *Talanta*, 58: 3–22.

Chapter 10

Physico-chemical Analysis of Some of the Selected Parameters of Sen Pokhar of Dumka (Jharkhand)

Sushmita Boyra, L.N. Patralekh, Saurabh Dutta and Prasanjit Mukherjee

ABSTRACT

Water, the most vital resource on this planet is adversely affected both qualitatively as well as quantitatively by various physical and chemical factors. The physical and chemical properties of water influence its suitability for specific application and also affect the general conditions of the aquatic environment and its biota.

Dumka district, (Jharkhand) is situated between 87°14′N–87°25′N latitude and 24°20′E–24°21′E longitude. It is situated on bank of river Mayurakshi. A pond commonly called Sen Pokhar is a large perennial pond of Nonihat of Dumka district. This Pond is fresh water with lentic as well as aquatic ecosystem and the same is used by the localities for their needs *viz.*, drinking, bathing, washing, fishing, boating etc. Physico-chemical analysis of this pond was done at monthly intervals during June 2009 to May 2010 and noticeable variations were recorded, some of which has found beyond the permissible limits.

Keywords: *Pond, Lentic, Algal mat, Abiotic factors.*

Introduction

Water is the most important natural resource, essential for all forms of life. Physically water is found in all three forms i.e. solid, liquid and gas. Pure water is liquid between 4°C –90°C, and heaviest at 4°C. It has highest specific heat, latent heat of fusion and latent heat of evaporation, than any other common substance. It is a common solvent. No other common substance can be compared with it in this respect. That is why it is called 'Universal solvent'. It allows dissolution of electrolytes, polar and non-polar electrolytes and even some non-polar chemicals. It has high surface tension and cohesiveness, due to loose Hydrogen bonds present between the water molecules.

Chemically water is a dihydride of Oxygen. It is composed of two molecules of Hydrogen and one molecule of Oxygen. Its molecular formula is H_2O. Both the Hydrogen atoms are attached to the Oxygen atom by co-valent bonds. The angle between the two Hydrogen atoms is 104.5°. The Oxygen atom has a partial negative charge ($ä^{-}$) while both the Hydrogen atoms have a partial positive charge ($ä^{+}$) (Nelson and Cox, 1982).

Ponds are small bodies of fresh water. These are relatively small and shallow, their littoral region is large and the limnetic and profundal regions are small or may even be absent. Ponds are most common and important reservoirs of fresh water, which are widely used by people for various purposes like bathing, washing, drinking, fishing, recreation, irrigation etc. Dumka comes under tropical climatic condition is divided into three distinct summer, rainy and winter season. 'Pokhar' are a type of large perennial pond. It is generally man made but may be natural also. They may be seasonal or perennial.

The present work "Physico-chemical analysis of some of the selected Parameters from Sen Pokhar of Dumka, Jharkhand"is an attempt to analyze physico-chemical factors of this Pokhar, as it is widely used by a large mass of people and has not yet been explored scientifically. It is the most important source of potable water, which plays a pivotal role in the development of society. It is widely used by the thousands of local inhabitants for bathing, washing, fishing, recreation, swimming, drinking, dumping of garbage and immersion of idols and images of Gods and Goddesses.

The pond water is blooming and stinking due to luxuriant growth of heavy algal mat on the surface of water. The pond appears to be polluted. Physico-chemical factors are important aspects of aquatic environment on the basis of which quality of water can be judged. Taking this view in mind the present work has been undertaken to assess some physico-chemical parameters of Sen Pokhar which reflected some valuable informations of ecological and environmental interests.

In initial investigations, stresses were given to study the physico-chemical characteristics of lotic ecosystem. Important contributions on fresh water lotic ecosystems were made by Gunnerson (1967), Pruthi (1933), Ganapati (1941-43), Michael (1964-69), Hussaini (1965), Kumaran (1966), Verma (1967), Qudari and Mustafa (1984), Goel *et al.* (1985), Kumar and Siddique (1997), Kumar and Verma (2001) etc.

Methodology

The sampling of the water of Sen Pokhar of Dumka district was done at monthly intervals from June 2009 to May 2010 for the study of physico-chemical properties. Along with the seasonal fluctuations, the diurnal variations in physico–chemical characteristics of the water were also studied. Monthly samples were taken from the surface and sub-surface water of the Pond. All the samplings were done in the morning hours between 8:00a.m. to 12:00noon, throughout the study period. Samples were properly collected in plastic bottles following, Golterman *et al.* (1978) and then brought to the laboratory for analysis. Estimation of Temperature, Turbidity, Total solids, Hydrogen ion Concentration (pH), Dissolved oxygen were done on spot at sampling site immediately, just after collection of water sample. Analysis of other parameters was done in the laboratory within three to four hours of the sample collection.

Standard methods were followed to analyses physico–chemical parameters. For analysis of Physico-chemical parameters, standard methods given in reputed work books of APHA, AWWA, FSIWA, NEERI, Adoni etc. were followed.

Results and Discussion

Water, the most vital resource for life on this planet which is adversely affected both qualitatively and quantitatively of physical, chemical factors. The physical and chemical properties of water influence its suitability for specific application and also affect the general conditions of the aquatic environment and its biota. Water is never found in its pure state in nature. Fresh water originates from atmospheric precipitation, which includes extremely variable quantities of many inorganic and organic substances. Once precipitation falls on the earth it runs and flows over soils and rocks. It dissolves various substances and may become highly mineralized. The degree of mineralization depends upon the time of contact between water and the composition and solubility of mineral matters. When the runoff passes through the agricultural areas, it dissolves various organic compounds. Finally, water reaches to a lentic and lotic environment with a varied amount of inorganic and organic substances.

Water quality is also affected by silt, organic and other allochthonous materials dissolved and suspended matter and addition of excesses of agricultural chemicals.Surface water is often used for the disposal of human waste, municipal sewage and industrial effluents, the inorganic and organic elements coming from these sources introduced into the water, which affects the delicately balanced functional systems. Besides this, acid rains also have a serious impact on the surface water. It is to be noted that, some of the gases, mineral constituents, soluble and suspended organic and inorganic matters are also influenced by many physical factors like atmospheric temperature, light, water current etc.

Besides these, biological processes also alter the features of water, through the uptake of ions or gases, or through the release of metabolites. Sometimes biological processes are considered as responsible factors for regulating the concentration of many substances, among them photosynthesis and respiration are mainly responsible for dissolved oxygen and carbon dioxide in natural waters. Further, many organic compounds are synthesized by living organisms of water affect the nature of the water.

The most important physico-chemical parameters affecting the features of water are transparency, light, temperature, pH, carbondioxide, carbonates, bicarbonate, nitrates, phosphates, hardness, chlorides, silicates and dissolved oxygen etc. Physical and chemical properties of water and the dependence of all life processes on these factors, make it desirable to take water as an environment.

After taking into consideration of all important factors and parameters, in the present investigation was under-taken, to understand seasonal variations as well as co-relation pattern between these parameters. Data were collected at monthly intervals covering all important factors. Results of different physico-chemical factors recorded from Sen Pokhar during June 2009 to May 2010 are as follows:

The parameter of temperature is basically important for its effect on the chemistry and biological reactions in the organisms present in water. A rise in temperature in the water leads to the spreading up of the chemical reactions, reduces the solubility of gases and amplifies the taste and odors (Trivedy and Goel, 1984).

The temperature of air at the research site ranges from 30°C to 32°C in rainy season, 16°C to 24°C in winter and 22°C to 33.4°C in summer. Higher temperature was recorded in summer season followed by rainy and winter season. Highest temperature was found in the month of June and lowest in February.

Water temperature follows the trend of variation in ambient temperature, but is influenced by substrate composition, insolation, turbidity, sewage effluent discharge and rain water inflows.It ranged between 32.2°C to 33°C during winter and 24°C to 33.8°C during winter (Table 10.1). Water temperature was found minimum in summer. Lowest water temperature (18.5°C) was recorded in the month of January 2003, while the highest in June. Slight downward trend was noticed during rains. Significant fluctuation in water temperature was recorded which is in conformity with the reports of Saha (1985), Saha and Pandit (1984). Direct impact of air temperature on water temperature was observed. Higher water temperature during summer might be due to direct solar radiation, low water level, clear atmosphere and longer duration of days.

pH is also very important parameter of limnological study. Most natural waters are generally alkaline due to presence of sufficient quantities of carbonates. pHof water gets drastically changed with time due to the exposure to air, biological activity and temperature changes. Significant changes in pH occur due to disposal of industrial waste, acid, mine, drainage etc. In natural waters, pH also changes diurnally and seasonally due to variation in photosynthetic activity, which increases the pH due to consumption of CO_2 in the process (Trivedy and Goel, 1984).

Table 10.1: Some Physico-Chemical Parameters of Sen Pokhar, Dumka

Parameters/Unit	*2009*							*2010*				
	Rainy Season				*Winter Season*				*Summer Season*			
	Jun	*July*	*Aug.*	*Sept.*	*Oct*	*Nov*	*Dec*	*Jan*	*Feb*	*Mar*	*Apr*	*May*
Air Temp (°C)	33.4	31.8	30.0	30.0	32.0	24.0	18.0	18.0	16.0	22.0	25.5	28.5
Water Temp (°C)	33.8	32.2	33.0	32.2	32.5	25.0	19.0	18.5	19.0	24.0	27.5	31.0
pH	7.3	6.9	7.1	6.8	6.5	6.2	5.8	6.2	6.5	6.9	7.2	7.3
DO (ppm)	4.7	7.4	6.6	2.4	2.8	2.3	1.0	0.4	1.1	3.0	3.5	5.8
Turbidity (NTU)	90.0	60.0	60.0	80.0	50.0	40.0	40.0	80.0	20.0	40.0	60.0	70.0
Conductivity	520.0	400.0	360.0	260.0	330.0	300.0	270.0	280.0	365.0	450.0	550.0	850.0
Dissolved Solids (ppm)	400.0	550.0	180.0	640.0	300.0	200.0	200.0	300.0	300.0	320.0	300.0	800.0
Suspended Solids (ppm)	50.0	260.0	380.0	620.0	110.0	110.0	70.0	310.0	180.0	110.0	140.0	80.0
Total Solid (T.S.)	450.0	810.0	560.0	1,260.0	410.0	310.0	270.0	610.0	480.0	430.0	440.0	880.0
Si (ppm)	25.8	16.1	10.0	13.1	17.2	16.1	17.0	14.1	12.7	14.2	21.7	36.8
Cl (ppm)	53.6	43.0	36.1	16.4	15.1	23.8	25.0	26.8	25.8	24.8	59.6	77.4
Total Hardness (ppm)	66.0	35.0	34.8	35.0	44.8	42.0	34.8	42.0	44.0	47.6	77.6	70.0
Ca (ppm)	60.0	31.0	24.6	30.0	31.2	36.0	32.0	36.0	42.0	40.0	68.0	66.0

Hydrogen ion concentration (pH) fluctuated in between 6.5 to 7.1 during rainy season, 5.8 to 6.5 in winter and 6.9 to 7.3 in summer (Table 10.1). Its lowest value was recorded in December and highest in June. Variation in pH value was narrow but irregular (Table 10.1). Variation in pH value remained near the juncture of acidic and alkaline border. It is assumed that pH is regulated by Carbon dioxide and Bicarbonate. Sen Pokhar water was neither strongly acidic nor strongly alkaline (Hutchinson, 1957).

Dissolved oxygen concentration in natural water is of primary importance as a regulator of metabolic process of plant and animal community and as an indicator of water quality. Dissolved oxygen concentration of water is constantly changing because of the biological, physical and chemical processes. The photosynthetic activity of aquatic vegetation is the major factor controlling the amount of dissolved oxygen in the water. Direct dissolution of oxygen due to surface contact with water is also a good source of DO in water bodies.

Dissolved oxygen (DO) is an important parameter to judge the quality of water. Oxygen is essentially in between 2.4 ppm to 7.4 ppm during rains, 0.4 ppm to 2.32 ppm in winter and 2.96 ppm to 5.84 ppm in summer months. Its lowest value (0.4 ppm) was observed in the month of January, 2003 and highest (7.4 ppm) in the month of July. Table 10.1 displays its monthly variation, which is very irregular. Its lower amount might be due to the standing state of water, least water current, greater decomposition and increased respiration by heterotrophic organism which is in conformity with the findings of Saha and Choudhary (1985). Its higher DO may be attributed to greater aeration by rippling, turbulence, water current and wind velocity. Higher rate photosynthesis due to greater phytoplankton population might be also the reason of higher DO in winter.

Turbidity is the status of water in respect of suspended particles in water. Suspended particles make the water turbid and opaque, this state of water is known as the turbidity. It depends on the amount of suspended particles present in water. Turbidity ranged between 50 to 80 Nepthelo Turbidity Unit (NTU) in rainy season, 20 to 80 NTU in winter and 40 to 90 in summer season (Table 10.1). Its lowest value was found in the month of February while the highest in June.

Conductivity is also a very important physical parameter of fresh water. It is a good and rapid measure of the total dissolved solids. Conductivity of an aquatic ecosystem shows the ability of electric current to flow more frequently through the water and depends upon the mineral content dissolved in the water and temperature. It is an important criterion in determining the suitability of water and wastewater for irrigation (Trivedy and Goel, 1984).

During the present investigation, marked variations were observed in the electrical conductivity at the site. In the investigation, conductivity value ranged from 260 to 400 during rainy season, 270 to 365 in winter and 450 to 850 in summer (Table 10.1). It showed a wide range of variation. Its lowest value was recorded in the month of September while, the highest in the month of May. Lower conductivity during rains was probably due to dilution of water by the addition of rainwater. Higher conductivity was found in summer months, which is attributed to depletion in the volume of water.

The total solid is the sum total of suspended solids and dissolved solids. The total solid is affected by the percentage of dissolved solids and suspended solids in water. It is also influenced by temperature, water level and amount of organic matters.

In Sen Pokhar pond total solids fluctuated in between 410 to 1260 ppm in rainy season, 210 to 610 in winter and 430 to 880 ppm in summer (Table 10.1). Its minimum value was recorded in the month of December while the maximum amount in September. Its higher amount was observed in rainy season and lower during winter. Dissolved solids ranged between 180 ppm to 640 ppm during rainy season, 800 to 300 ppm in winter and 300 to 800 ppm in summer. Suspended solide varied from 110 to 620 ppm in rainy season, 70 to 310 ppm in winter and 50 to 140 ppm in summer (Table 10.1). Higher amount of total solids in rainy season was perhaps due to the addition of muddy rain water having grater amount of suspended organic matter and silt particles.

Silicate is the most abundant element of the earth after oxygen. It is found in the form of Silicate in natural water. Despite, its over abundance in nature, it occurs in meagre quantities in fresh water. This is due to its resistant nature against the chemical weathering process (Trivedy and Goel, 1984). Presence of Silicate is very essential for the frustules formation in diatoms. In the present investigation the silicate concentration has clearly depicted its seasonality.

During the period of investigation, the concentration of silicate varied between 10.00 ppm to 17.20 ppm during monsoon months, 12.69 to 17.00 ppm in winter and 14.2 to 36.76 ppm in summer. Its lowest amount was analysed in the month of August while the highest in May. It exhibited a wide range of variations. Comparatively its highest values were recorded during summer months followed by winter and monsoon.

Chloride occurs naturally in all types of waters. In natural fresh waters, however, its concentration remains quite low and is generally less than that of sulphate and bicarbonates.

Excreta of animals and human origin, which contains large quantities of chlorides together with nitrogenous compound and their disposal in the natural waters, are the major sources of chloride in fresh water. Generally its high concentration indicates pollution in water which may be of domestic sewage origin.

Chloride metabolically depicts pivotal role in photolysis of water and phosphorylation in autotrophs. It is considered as pollution indicator when present in high concentration. Table-1 reveals the Chloride concentration of different months. It fluctuated in between 15.08 to 243.5 during rainy season, 23.82 to 26.8 ppm in summer. The range of its variation was wide during monsoon months, narrow in winter and wider in summer. Its lowest amount was recorded in October while highest in May.

Calcium is one of the most abundant micro nutrients of the natural water. Natural water usually contains Calcium salts. Concentration of Calcium is regulated by several factors such as, leaching from rocks, dissolution of Calcium salts and their reduction at high pH. Concentration of the Calcium is reduced at higher pH due to it's precipitation as $CaCO_3$.

The domestic sewage and industrial effluents are good source of Calcium ions, so their disposal increases the Calcium concentrations in water.

During the two years of analysis the value of Calcium ranged from 24.6mg/L in August to 68.0 mg/L in April.

Hardness is the property of water which prevents the lather formation with soap and increases the boiling point of waters. Principal cations imparting hardness are calcium and magnesium. However, other cations such as Strontium, Iron and Manganese also contribute to the hardness. The anions responsible for hardness are mainly bicarbonate, carbonate, sulphate, chloride, nitrate and silicates etc. Hardness is called temporary if it is caused by bicarbonate and carbonate salts of the cations, since it can be removed simply by boiling the water. Permanent hardness is caused mainly by sulphate and chlorides of the metals (Trivedy and Goel, 1984).

During the two years of study period the hardness of water fluctuated within a range of 31.90mg/L (site A in July, 05) to 91.60mg/L (site C in May, 05). The average hardness of water of the bandh in two years, was 54.45mg/L.

During the study period the hardness of water fluctuated within a range of 34.8 to 44.8 ppm in winter and 47.6 ppm in summer. Its lowest value was observed in the months of August and December while, the highest in April. Addition of rainwater during monsoon might be the reason of its lower value in rainy season.

The monthly variation in the amount of the Calcium hardness which ranged between 24.6 to 31.2 ppm during monsoon months, 32.00 to 42.00 ppm in winter months and 40.00 to 68.00 ppm in summer months. It showed narrow range of variation during rainy and winter season while, a bit wide range in summer season. Its lowest value was recorded in September and highest in the month of April. Nutrients, current and human activities impart on the amount of Calcium hardness (George, 1966). Lower amount of calcium hardness in September might be caused by greater dilution due to addition of rainwater, which supports the findings of Singh (1985).

References

1. APHA. (1975). *Standard Methods for the Examination of Water and Wastewater*, 14th edn. American Water Works Association, Washington, New York, pp. 1193.

2. Adoni, A.D. (1975). Studies on microbiology of Sagar Lake. *Ph.D. Thesis*, Univ. of Sagar (M.P) India, 254 pp.

3. APHA, AWWA, WPCA (1998). *Standard Methods for the Examination of Water and Wastewater*, 20th Edn. American Public Hlth. Assoc., Washington, USA.

4. Ganapati, S.V. (1941). Studies on the chemistry and biology of ponds in Madras city-seasonal change in the physical–

chemical conditions of a garden pond containing aquatic vegetation. *J. Madras Univ.*, 13(1): 55–59.

5. Ganapati, S.V. (1943). An ecological survey of a garden pond containing abundant zooplankton. *Proc. Indian Acad. Sci.* (B), 17: 41–58.
6. Goel, P.K., Trivedi, R.K. and Bhave, S.V. (1985). Studies on the limnology of a few fresh water bodies in South Western Maharastra. *Indian J. Env. Prot.*, 5(1): 19–25.
7. Golterman, H.L., Clymo, R.S. and Ohnstad, M.A.N. (1987). *Methods for Physical and Chemical Analysis of Freshwaters*, 2nd edn. I.B.P. Manual. Blackwell Scientific Publ., Oxford, UK.
8. Hussainy, S.U. (1965). Limnological studies of the departmental pond of Annamalainagar. *Environ. Health*, 7: 24–37.
9. Kumar, A. and Siddiqui, E.N. (1997). Quality of drinking water in and around Ranchi. *Indian J. Environ. Prot.*, 18(5): 339–345.
10. Kumar, A. and Verma, P.K. (2001). Ecological status of Masanjore Reservoir in relation to fisheries management in Santhal Pargana (Jharkhand), India. In: *Ecol. and Conservation of Lakes, Reservoirs and Rivers*, (Ed.) A. Kumar. ABD Publishers, Jaipur, pp. 889.
11. Kumaran, P. (1966). A natural sewage stabilization pond at Jaipur. *Environmental Health*, 8: 134–141.
12. Michael, R.G. (1964). Diurnal variation of the plankton correlated with physico-chemical factors in three different ponds. *Ph.D. Thesis*, Calcutta University, Calcutta.
13. Michael, R.G. (1969). Seasonal trends in physico-chemical factors and plankton of fresh water fish pond and their role in fish culture. *Hydrobiologia*, 33(1): 144–160.
14. NEERI (1986). *Manual on Water and Wastewater Analysis.* National Environmental Engineering Research Institute, Nagpur, pp. 340.
15. Pruthi, H.S. (1933). Studies on, the, bionomics of freshwaters in India. I. Seasonal changes in the physical and chemical conditions of the waters of the tank in the Indian Museum compound. *Int. Rev. Ges. Hydrobiol. Hydrogr.*, 28: 46–47.

16. Saha, L.C. (1985). Changes in the properties of bottom soil of two freshwater ponds in relation to ecological factors. *Indian J. Ecol.*, 12(1): 147–150.

17. Saha, L.C. and Pandit, B. (1984). Comparative ecology of Bhagalpur ponds and river Ganges during summers. *Nat. Acad. Sci. Letters*, 7(10): 295–296.

18. Trivedi, R.K. and Goel, P.K. (1986). *Chemical and Biological Method for Water Pollution Studies*. Environment Publications, Karad, India.

19. Verma, M.N. (1967). Diurnal variation in a fish pond in Seoni, India. *Hydrobiologia*, 30(1): 129–137.

Chapter 11

Techniques for Providing Good Quality of Water to the Society

K. Swarnim, Nilima Verma, Neetu and R.K. Singh

ABSTRACT

An ideal waterworks management should ensure that the water supply for the public distribution should be free from pathogenic organisms, undesirable taste and odour, should be clear, palatable, of reasonable temperature neither corrosive nor scale-forming in pipes and free from minerals which could produce undesirable physiological effect.

Introduction

Rain water is the purest form of water. But when it comes to ground, it carries dust and dissolves certain gases forming acid. It collects decayed matters from vegetables etc. which forms organic acid in water. When water is supplied to a town from public water supply scheme, it should be made free from harmful impurities, salts and bacteria. Water may have salt and mineral but in limited amount so that they give good taste and also assist in digestion. Polluted water causes various waterborne diseases like cholera, dysentery, jaundice etc. So water should be tested and treated before distribution in public.

Analysis of Water

Testing of water is done in laboratory to confirm the standard of water supplied to consumers. These tests are conducted[1–5]:

1. To determine the quality of raw water.
2. To determine the treatment process to be provided.
3. To ensure that treatment of water is properly done during each phase of treatments.
4. To examine whether the treated water conforms to standards.

Type of Tests

Three types of tests are conducted to determine the quality of water:

a. Physical Test

It indicates the aesthetic quality and performance of various treatment units.

b. Chemical Test

It indicates the amount of chemicals present in water.

c. Bacteriological Test

It indicates the presence of bacteria producing pollution and whether water is safe for drinking purpose or not.

Sampling

Sampling of the collected water is done in different steps. Sample of water collected should have date and time of collection, source of sample, and temperature of water at the time of collection must be noted. Sample is collected in polythene of Pyrex glass. Glass bottle may be rinsed with chromic acid before used. Bottles should be cleaned before used. They are packed in wooden, metal, plastic or heavy fiber board cases. About 2-5 liter of sample is required for analysis. The sample should reach the place of analysis within 72 hours of collection.

Physical Tests

Physical test of water includes the following analysis:

a) Colour

The colour of sample may be due to presence of fine particles or certain minerals in solution. Colour is measured with tintometer. The unit of colour is measured on platinum-cobalt scale.

b) Taste and Odour

Taste and odour in water is due to the presence of one or any of the following:

1. Micro organisms either dead or alive.
2. Dissolved gas such as CO_2, O_2, CH_4, H_2S with organic matter.
3. Minerals substances as NaCl, Fe components, Carbonates and sulphates of other elements.

Taste may also originate in the treatment process due to excess chlorination. Odour of the water changes with temperature and are classified as fishy, grassy, earthy, vegetable, sweetish etc. It is tested with 'osmoscope' at 20-25°C.

c) Temperature

Temperature has no significance because practically nothing can be done to reduce temperature.

d) Turbidity

Turbidity of water is due to the presence of visible material or suspended inorganic matter like silt, clay etc. It is expressed in ppm by weight of suspended matter in water.

Chemical Tests

Chemical test of water is carried out to determine:

a) Total Solids

1. Dissolved solid
2. Suspended solid

Suspended solids are formed by filtering the water through fine filter. The material in filter paper is then weighed. The filtered water is then evaporated and residue is weighed. This gives the dissolved matter.

b) Hardness

Hard-water is those which do not lather with soap. It is of two kinds:

1. *Temporary hard water*: It is due to the presence of bicarbonates of Ca or Mg.
2. *Permanent hard water*: It is due to the presence of sulphates, chlorides and nitrates of calcium and magnesium.

Hardness of water is tested by EDTA. Here water is titrated against EDTA salt solution using Eriochrome BlackT as indicator. The colour changes from red to blue while titrating.

Hardness of water is expressed in ppm of $CaCO_3$ in 1 liter of water.

Chlorides

The amount of NaCl present in water is determined by adding $AgNo_3$ of known concentration and Potassium chloride to the water to be tested. The solution is continuously stirred. If chloride is present then reddish colour will be formed. Permissible chloride is 250 ppm.

Dissolved Gases

Oxygen dissolved in surface water is termed as dissolved oxygen. It depends upon the amount and quantity of unstable organic matter. (H_2S, CH_4, CO_2, Cl are formed dissolved rarely). This DO of water is measured by exposing sample of water for 4 hours at 27°C with $KMnO_4$ of 10 per cent concentration. The oxygen dissolved should be between 5 and 10 ppm.

pH Value Hydrogen-ion Concentration

This test is conducted to find the acidity or alkanity of a sample of water. Alkanity in water is caused by bicarbonates or hydroxides of Na, K, Ca and Mg.

When electric charge is passed through water, it dissociates into positively charged and negatively charged ions.

pH value is determined by following two methods:

1. Electrometric method
2. Colourimetric method.

Nitrogen and its Compound

Nitrogen is present in water in following forms:

1. Free ammonia
2. Nitrites
3. Nitrates

Fluorides

These occurs chemical waste from industries. Fluorides are beneficial in water if present in small concentration up to 1 ppm.Such fluoride improve dental health and prevent dental caries.

Metal and Other Chemical Substances

Iron present in water above a certain limit imparts taste to the water.Manganese causes organic growth[6–7].

It blocks pipes and discolors clothes. This should not exceed 0.05 ppm.The amount of Iron, Manganese and other metal in water are determined by adding coloring agent to the sample of water and comparing with solutions of known amount of metals.

Bacteriological Test

Water may contain bacteria which are very small organism and can be seen only under microscope. Following test are done for bacteriological examination of water:

1. Total count test.
2. *E. coli* test or Coliform test.

Most Probable Number (MPN)

When a number of tests are conducted on portions of a sample, the value obtained in most of the test give bacterial density *i.e.* bacteria content which is most likely to the present in water, and is termed as most probable number (MPN).

If on examination all portions are negative, the MPN is 0. If only 1is positive value is 2.2 per 100 ml and so on.

References

1. Ewing, G.W. (1969). *Instrumental Method of Chemical Analysis*, 3rd Edn. Mc Graw-Hill, New York.

2. Willard, H.H., Merritt, L.L. and Dean, J.A. (1974). *Instrumental Methods of Analysis*, 5th Edn. Van Nostrand, New York.

3. Skoog, D.A. and Leary, J.J. (1992). *Principles of Instrumental Analysis*, 4th Edn. Saunders College Publishing, Fort Worth.

4. Day, R.A. Jn. and Underwood, A.L. (1991). *Quantitalive Analysis*, 6th Edn. Prentice-Hall, Englewood Cliffs, NJ.

5. Skoog, D.A.M. West and Holler, F.J. (1988). *Fundamentals of Analytical Chemistry*, 5th Edn. Saunders College Publishing, New York.

6. Pontions, F.W. (1992). Acuurent look at the federal drinking water Regulation. *J. AWWA*, 84(3): 36.

7. Pontius, F.W. (1993). Federal drinking water regulation update *J. Aswin*, 85(2): 42.

Chapter 12

Acute Cytogenetic Toxicity by Distellery Influent in *Allium cepa* and Bone Marrow Cells of Mice

K.S. Awasthy, Satya Prakash,
G.P. Sinha and Nutan Kumari

ABSTRACT

Untreated wastewater (0.05, 0.5, and 1.0 per cent influent) from Bhagalpur based distillery, induced acute cytogenetic toxicity in mitotically dividing *Allium cepa* root tips and bone marrow cells of mice. The observations suggest that influent discharged carry mutagenic derivatives causing damages to crops and aquatic lives.

Keywords: *Influent, Genotoxicity, Aneuploidy, Mutagen, Allium cepa.*

Introduction

Water is a precious gift of nature. It is essential for life. Though widespread on the earth, hardly 3 per cent is available for drinking and irrigation purposes. The demand of fresh water is increasing day by day with the exponential growth of population, industry and agriculture. Therefore clean water is very essential for the live to survive on and it deserves immediate attention of ours from existence point of view. However, pollution of water due to indiscriminate use and disposal of wastes has attained serious dimensions worldwide (Bhasale, 1985). The release of toxic substances into the water bodies mostly from townships and industries is a major cause of water pollution. There are several reports of phytotoxic effects resulting from irrigation of crops with effluents discharged from different industries (Somashekar *et al.*, 1992, Hariom and Arya, 1994, Nath *et al.*, 2007, and Sahu *et al.*, 2007). The present study has been undertaken to evaluate the mutagenic (pollution) potential of the influents (Untreated wastewater) from the Shiv Shanker Chemical Ltd. Industries, Jagdishpur, Bhagalpur on the mitotically dividing chromosomes of *Allium cepa* root tips and bone marrow cells of mice *Mus musculus*.

Materials and Methods

Samples of influent from SCL, Bhagalpur (Bihar,India) were collected around 8.00 am for seven consecutive days during the summer season (1st week of june). The collected samples were observed for color and odor analyzed for pH. The collection, storage and analysis of samples were performed as per methods given by APHA (2005). The collected samples were pooled together to make a composite sample for the evaluation purposes and a part of it was diluted (u/v) with glass double distilled (gdd) water in to 0.05(lower), 0.5 (middle), and 1.0 (Highher) percentages. The cytogenetic effects of the diluted influents were studied on the chromosomes of mitotically dividing cells of *Allium cepa* (Onion) root tips and bone marrow cells of mice *Mus musculus*.

1. Plant Test System (Allium Test)

Since the root tips of plants come into direct contact with the water and the Allium test is simple, reliable, easy besides being in expensive, the test was carried out to assess the direct effect of influent on chromosomal aberrations (2n = 16). Healthy onion bulbs weighting 50±5g were grown on gdd water for 3-4 days. One set of eight bulbs were grown on fresh water as control and three sets (T_1, T_2 and T_3) comprising eight bulbs each were transferred to different concentrations of the influent after 24 hrs.

After 24 hrs of exposure, roots were thoroughly washed in tap water and last few mm long root tips were cut and directly placed into the solution of para dichloro-benzene for 2 hrs at 4-10°C. They were washed and transferred to carnoy's fixative (ethyl alcohol/ glacial acetic acid, 3:1 v/v) for 24 hrs at 4°C temp. Finally they were preserved in 70 per cent alcohol (4-10° C). Slides of tip cells were prepared by squash technique (Fiskesjo, 1981 and 1982) after transferring the root-tips into HCl-C_2H_5OH misture (5mts) acetocarmine solution (1 hr) and rinsed in 45 per cent acetic acid. About 75-80 slides per experimental set (@ 10 slides/bulb) were prepared and screened @ 25-30 metaphase plates per slide. Abnormalities were then categorized as structural and numerical ones. Disturbed cytokinesis (binucleate/multinucleate) due to failure of cell-plate formation during cell division were also searched. The effect of influents on eukaryotic plant cell nuclei was also assessed by observing 2-3 thousand dividing cells and interruption in cell-division, *i.e.* Mitotic Index.

Table 12.1: Experimental Groups and Treatment Protocol

Test System	*Expt. Groups*	*Treatment Duration*	*Doses*
A.	Control (C)	24 hrs	Fresh water
Allium cepa	T_1 (Treated)	24 hrs	0.05 per cent
	T_2 (Treated)	24 hrs	0.50 per cent
	T_3 (Treated)	24 hrs	100.00 per cent
B.	Control (C)	7 days	@20ml/kg.bw/day.gdd
Mus musculus	T_1 (Treated)	7 days	0.05 per cent
	T_2 (Treated)	7 days	0.50 per cent
	T_3 (Treated)	7 days	100.00 per cent

2. Animal Test System

Experiments were carried out in 6-8 week old, body weight (bw) 25g laboratory inbred stock of swiss *albino* mice (Mus musculus, cdri-s, 2n=40). The mice matched for their age and sex were separated into four groups of six animals each for control and different concentrations of the influent. The control group recived gdd water @ 20ml/kg body weight (bw) per day while rest of the thre groups (T_1 Lower, T_2 Middle, T_3 Higher) were given different concentrations of the influent at the same rate. All the experimental animals were kept on Gulmohar diet (Hindustan Liver, Bombay) with provision of drinking water ad libitum. Treatments were continued for one week, once daily, by oral canula and mice were sacrified on 8th day by cervical dislocation. Colchicine (4 per cent) was injected

Table 12.2: Effect of Influents on Mitotic Index and Frequency (per cent ±SE) of Different Phases in Root Tip Cells

Expt. Groups	No. of Cells Observed	No. of Dividing Cells	Phase-wise Distribution of Dividing Cells			
			Pro	Meta	Ana	Telo
C (Control)	2250	1110	640±0.95	201±0.60	140±0.50	129±0.48
T_1 (Treated)	2010	995	623±1.03	193±0.65	166 ±0.61	123±0.53
T_2 (Treated)	2751	915	487±0.72	225±0.52	104±0.36	97±0.34
T_3 (Treated)	3014	914	466±0.65	220±0.47	137±0.37	91±0.31

Table 12.3: Frequencies (per cent ±SE) of Abnormalities in Chromosomes in Onion Root Tip Cells Induced by Influents @ 300 Metaphase/Group

Expt Groups (No.of Metaphase)	Chromosomal Aberrations							
	Structural Anaphase No.%±SE	Laggard No.%±SE	Numerical Polyplody		Disturbed Cytokinesis		Gross Abnormalities	
			No.	%±SE	No.	%±SE	No.	%±SE
C (301)	Nil	Nil	3	0.99±0.57	Nil		3	0.99±0.57
T_1 (300)	Nil	Nil	4	1.33±0.67	Nil		4	01.33±0.67
T_2 (311)	10.32±0.33	Nil	6	1.93±0.78	3	0.97±0.55	10	3.22±1.00
T_3 (305)	10.32±0.33	10.32±0.33	6	1.96±0.8	5	1.64±0.73	13	4.26±1.11

Table 12.4: Frequencies (per cent ±SE) of Abnormalities in Chromosomes of Bone Marrow Cells Induced by Influents @ 200 Metaphase/Group

Expt Groups (No. of Metaphase)	*Structural*		*No.*	*%±SE*	*Numerical Polyplody*			*No. %±SE*	*Total No. %±SE*
	Ctp	*af*			*An*	*Poly*	*Cmit*		
C (212)	2		20.94	±0.66	2	2		4 1.89±0.94	6 2.83±1.14
T_1 (231)	1		10.43	±0.46	2	3		5 2.17±0.95	6 2.59±1.04
T_2 (207)	3	2	62.89	±1.16	1	5	2	8 3.87±1.33	14 6.76±1.65
T_3 (220)	4	6	92.09	±1.32	3	4	4	11 5.0±1.45	20 9.09±1.93

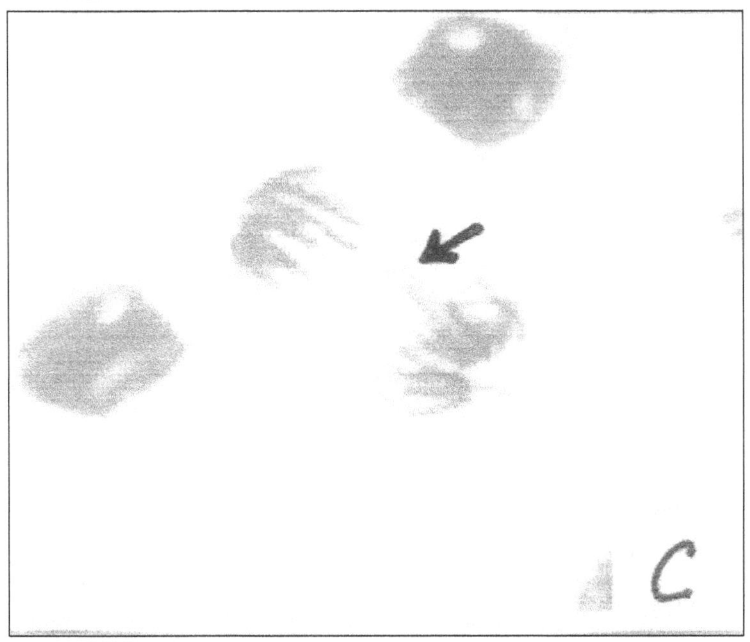

Plate 12.1: Anaphase Bridge

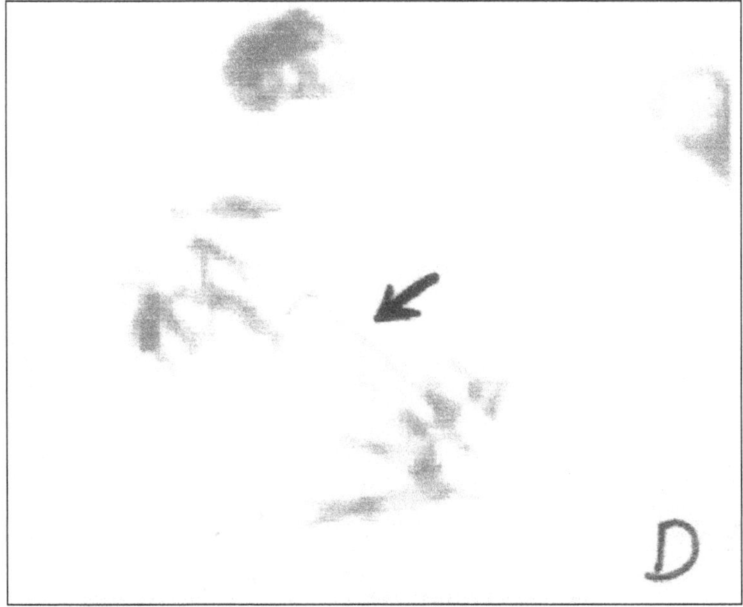

Plate 12.2: Anaphase with Laggard

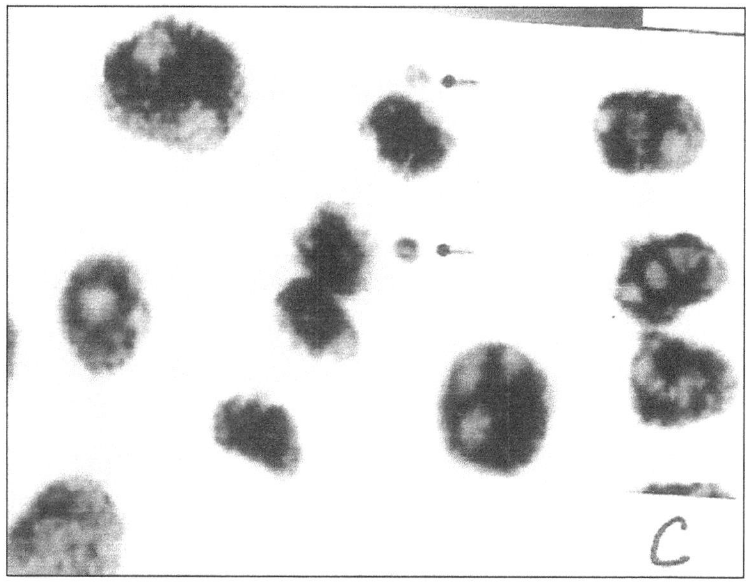

Plate 12.3: Polynucleate

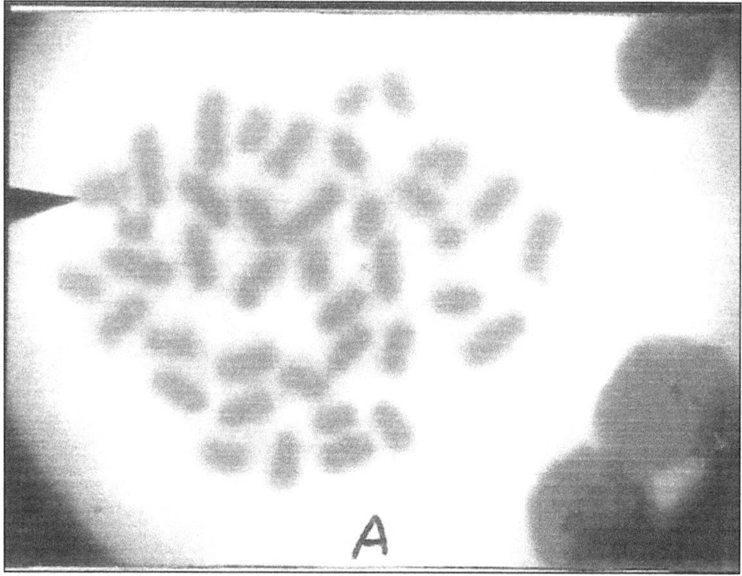

Plate 12.4: Normal Metaphase with 40 Well Spread Chromosomes

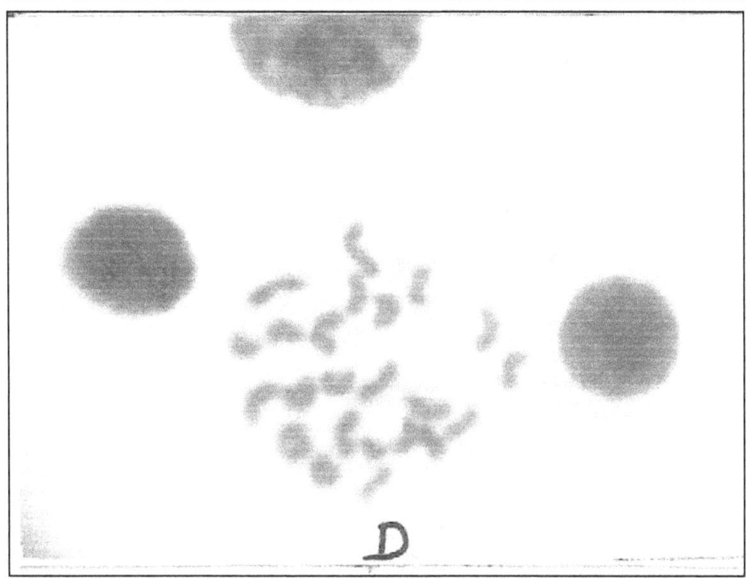

Plate 12.5: Aneuploidy

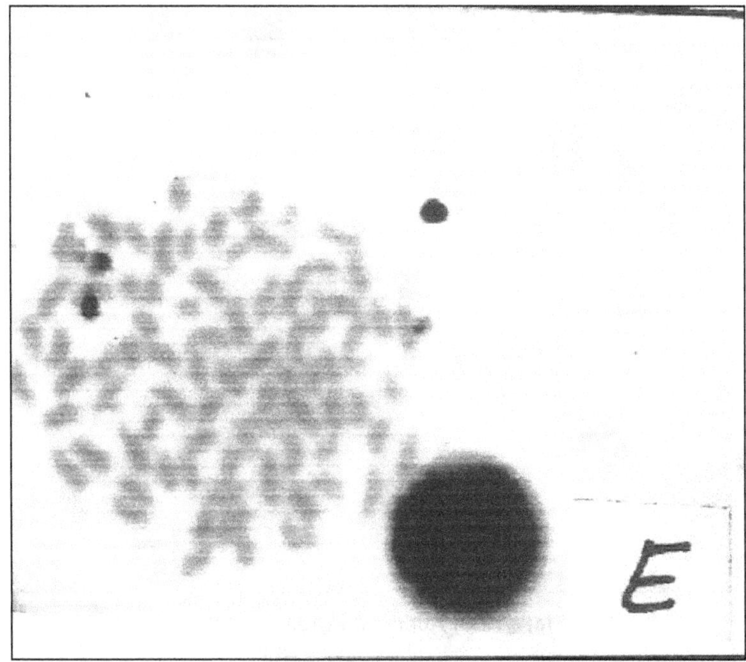

Plate 12.6: C Mitosis

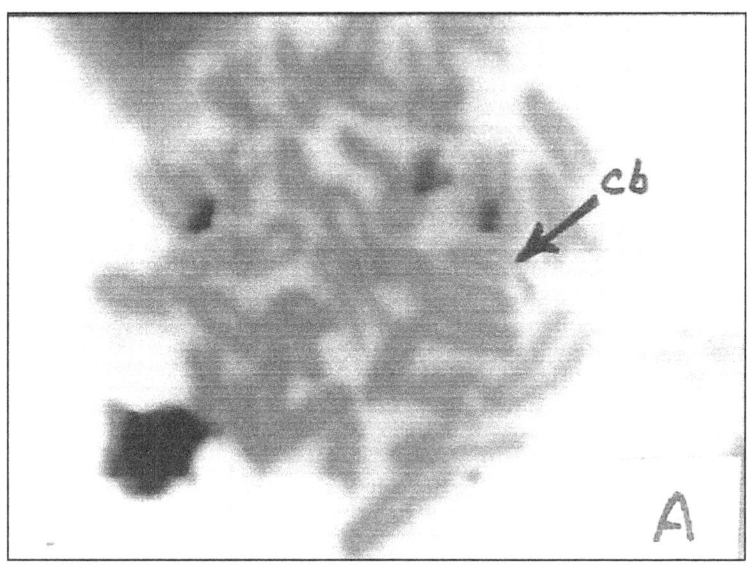

Plate 12.7: Chromosome Breaks

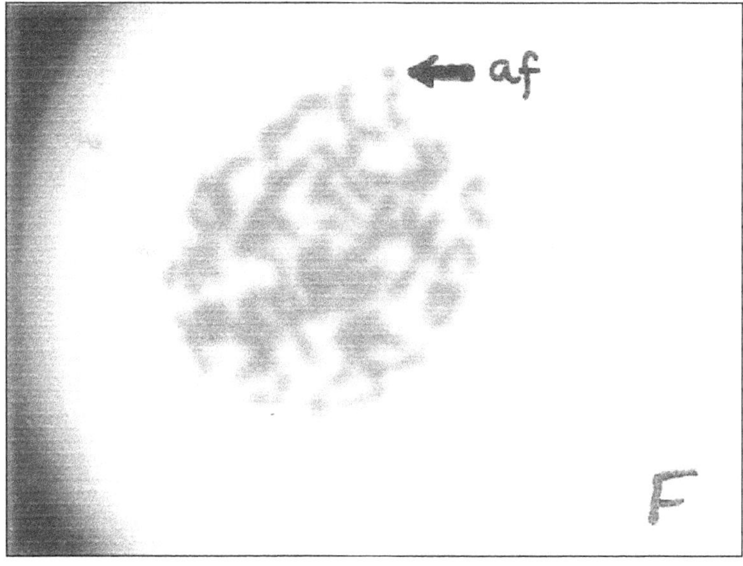

Plate 12.8: Acentric Fragment (af)

intraperitoneally @ 4ml/kg bw to harvest cells arrested at metaphase 90mts prior to sacrifice. Bone marrow was flushe out from both femora in prewarmed (37ÚC) KCl hypotonic solution. The slides were prepared by the fixative 3:1 and 1:1 aceto-alcohol – flame drying – Giemsa staining technique by Preston *et al.*, 1987. Atlest 300 well spread metaphase plates from each experimental group were screened randomly by selecting 50 plates per animal. The chromosomal abnormalities obtained were put into structural and mitosis disruptive categories.

For each treatment the experiment were replicated thrice and data were pooled for statistical analysis or proportion test (Downei and Heath 1970).

Results

Exposure of onion root tip cells with influent significantly ($P < 0.001$) reduced the MI. Phase wise analysis showed that the effect was harmful to every stages of the cell division regarding the chromosomal abnormalities in onion root tip cells the numerical changes were significantly high (1.96 per cent) in comparison to the control (0.99 per cent) at 100 per cent concentration. However numerical changes at this concentration were insignificant. While the gross abnormality was highly significant at 50 per cent and 100 per cent concentrations.

Discussion

The influent might be carrying heavy metals (Pandey *et al.*, 2008) which may cause mutations in onion root tip and bone marrow cells. Influent be treated seriously before discharge.

References

1. APHA (2005). *Standard Methods for the Examination of Water and Wastewater*, 21st Edn. American Public Health Association. Washington DC.

2. Awasthi, K.S., Chaurasia, O.P. and Sinha, S.P (1998). Prolonged murine genotoxic effects of crude extracted from neem. *Phytotherapy Res.*, 12: 1–3.

3. Bhosale, L.J. (1985). Effect of water pollution on plants. In: *Current Pollution Researches in India*, (Eds.) R.K Tiwary and P.K Goel. Environmental Publication, India.

4. Downeli, N.M and Heath, R.E (1970). In: *Basic Statistical Methods*, 3rd Edn. Harper and Row Publisher, Inc., New York, p. 86–239.

5. Fiskesjo, G. (1981 and 1982). Alliumtest on copper in drinking water. *Vatten*, 17(3): 232–240.

6. Hariom, N.S. and Arya, M.S. (1994). Combined effect of wastes of distillery and sugar mill on seed germination, seed growth and biomass of okra (*Ablemoschus esculentus*). *J. Environ. Pollut.*, 15: 171–175.

7. Nath, K., Singh, D. and Sharma, Y.K. (2007). Combinatonal effect of distillery and sugar factory effluents in crop plants. *J. Environ. Biol.*, 28: 577–582.

8. Pandey, S.N, Nautriyal, B.D. and Sharma, C.P. (2008). Pollution level in distillery effluent and its phytotoxic effect on seed germination and early growth of maize and rice. *J. Environ. Biol.*, 29: 267–270.

9. Preston, R.J., Dean, B.J., Galloway, S., Holdon, H., McFee, A.F. and Shelby, M. (1987). Mammalian *in vivo* cytogenetic assay Analysis of chromosome aberration in bone marrow cells. *Mutation Res.*, 189: 157–165.

10. Sahu, R.K., Katiyar, S., Tiwary, J. and Kisku, G.C. (2007). Assessment of drain water reciving effluent from tanneries and its impact on soil and plants with particular emphasis on bioaccumulation of heavy metals. *J. Environ. Bio.*, 28: 685–690.

11. Somashekar, R.K., Ramaioh, S. and Laxminayar, R.L. (1992). Effect of *Viqna sinensis* L and *Triqonella foenum-graecum* L. *J. Indian Biol. Soc.*, 71: 115–118.

Chapter 13

Water Pollution: Its Causes, Impacts and Control

Vandana Sahay

Introduction

The Latin word 'pollution' means defilement or the physical contamination of terrestrial or aquatic environment. Water pollution may be defined as the addition of any foreign material (*i.e.* inorganic, biological or radiological) or any physical change in the natural water which may harmfully affect the living life or other desirable species, our industrial process, living condition and cultural asserts directly or indirectly, immediately, after sometime or after a long time.

In India, a handful people get clean potable water and the rest quench their thirst from polluted water bodies.

Causes of Water Pollution

Water is polluted due to chemical pollutant and thermal pollutant.

The major chemical pollutants are micro-organism, organic wastes, plant-nutrients, toxic heavy metals, sediments oil and grease, agriculture run-off and radioactive substance.

The main source of chemical pollutants are domestic sewage, animal excreta and waste. Decaying animals and plants, erosion of soil, different industrial wastes, wrecked ships and ferries, mining etc. which leads to the depletion of oxygen called Eutrophication.

In thermal power plant, water is used as a coolant drawn from nearby water bodies. Due to difference in temperature the aquatic life is disturbed and the oxygen goes into the atmosphere and leads to Eutrophication (Figure 13.1).

It is very difficult to survive biomass of the aquatic ecosystem.

Recently on 15th March, 2011 a potential catastrophe explosion at stricken Fukushima Daiichi plant in Japan led to nearly 400 milli sieverts an hour. In CT Scan, the adult receives about 15mSv radiation while a new born baby receive 30mSv. Chest X-Ray involves about 0.02mSv while dental is 0.01mSv exposure. All living things died after completing their life cycle and release amino acids, phosphate etc.

The sulphate in water is the principal available which was reduced by autotrophic plants and incorporated in proteins (sulphure is an essential constituent of amino acids *i.e.* biomass) which is decomposed by heterotrophic micro-organisms, H_2S is released and then reconverted to sulphate by chemosynthetic organisms. Green Sulpher bacteria oxidized H_2S into 2S. H_2S may pass into reservisor pool where back and forth fluxing is very flow and Iron Sulphide, Sulphide Compound is formed and phosphorus dissolved in water and available for living being.

The mushrooming growth of industries increases poisonous gases in the environment which are responsible for lowering of water table, crisis of potable water, rising of sea-water-level, melting of glaciers, perennial river like Yammuna turns into nullah, production of crops may be reduced, decrease rate of photosynthesis, Green House effect, Global warming etc. which rapture the total aquatic plants and animals.

Ganga river is highly polluted in Kanpur, Allahabad and Varanasi. As small traces of radon *i.e.* a compound of radium is found in Ganga Water.

Yammuna river is heavily polluted at Delhi and Baghpat (Meerut). It is biologically dead river. It has been completely drying up in summer in Agra. The ebony shafts of TAJ MAHAL will not last long without constant seepage of water supplied by river Yamuna. The colour of monument is also affected due to acid rain.

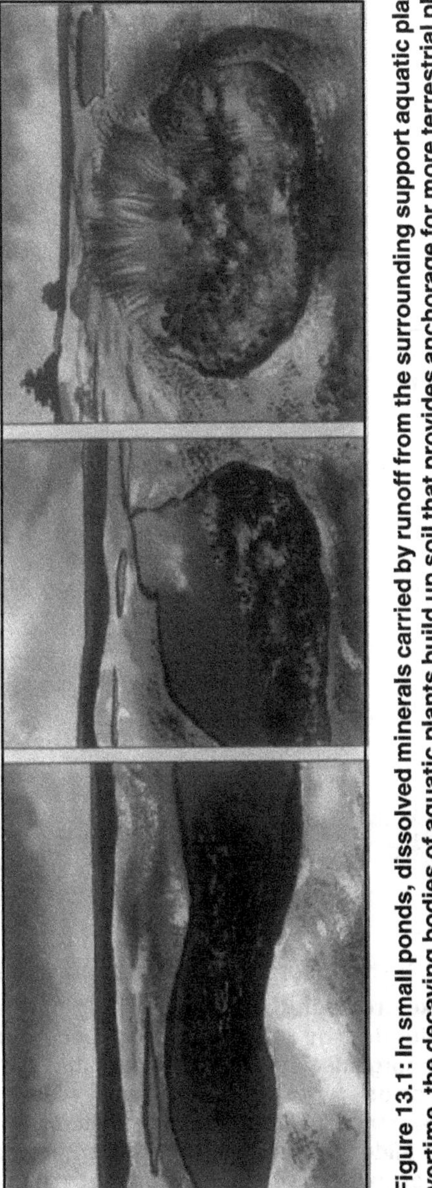

Figure 13.1: In small ponds, dissolved minerals carried by runoff from the surrounding support aquatic plants. Overtime, the decaying bodies of aquatic plants build up soil that provides anchorage for more terrestrial plants. Finally, the pond is entirely converted to dry land.

Impacts on Animals

The industrial waste like Cadmium, Mercury, Lead, Chromium, Nickel etc. expose hazards effect on human being and affects various organs such as liver, brain, kidney, lungs etc. and causes cancer in various parts of the body. Enteric diseases, skin disease, baldness, infertility, hypertension etc. are cause due to industrial wastes.

The administration of drug like Diclofenac to the cattle may proved fatal to the vulture, a natural sweeper and they are vanished from the environment whereas administration of Meloxicam is safe for vulture. Since amphibians are very sensitive to temperature. They also serves as indicators of climate change.

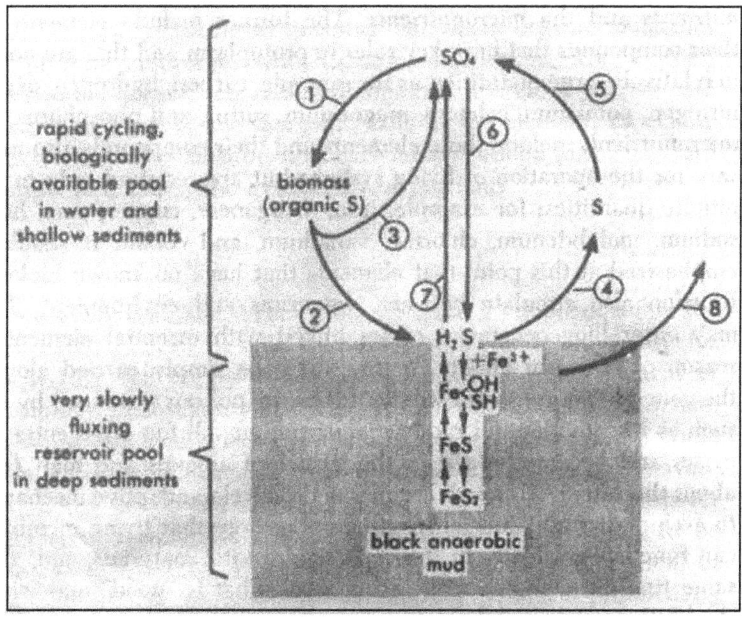

Figure 13.2: The Sulfur Cycle in an Aquatic System

Organisms play key roles in the rapidly cycling pool as follows: (1) Primary production by autotrophs; (2) Decomposition by heterotrophic microorganisms; (3) Animal excretion; (4), (5) Steps by specialized colorless, purple and green sulfur bacteria; (6) Desulfovibrio bacteria (anaerobic sulfate reducers); (7) Thiobacilli bacteria (aerobic sulfide oxidizers. Step 8 releases phosphorus (from insoluble ferric phosphate), thus speeding up the cycling of this vital element.

Radioactive substance in water causes eye cataract, blood cancer, mutation in genes and DNA breakage.

Methods to Control Water Pollution

The simple, inexpensive and micro technique to determine and remove the pollutant is categorized under Nanotechnology. These includes reverse osmosis, carbonnano tubes, nano-filtration, ultra filtration membrane etc.

Reverse technology is used to purify wastewater from sewage.

Bed of Alumunium turning can remove phosphorus compound from Match and chemical factories.

Chlorination is used to disinfectant drinking water and wastewater inhibit bacterial action and destroys objectionable tastes and odour.

Water hyacinth is a natural filter to absorb toxic substance from industrial effluents and domestic sewage.

The photolytic reaction in presence of the catalyst Titanium dioxide destroys bacteria, cyanides, and other poisonous chemicals

Researcher's of Belfast's queen's university have proposed peat filters from the root of cotton plant which can effectively and quickly absorb in Lead, Mercury, Cadmium and Chromium and peats can be washed with acid to remove metals.

Red mud is used to remove Cd. The barks of babul adsorb toxic metals like chromium, cadmium, lead, mercury, cobalt etc.

Recently V.P. Kudesia and his co-worker discovered technique from the bark of babul, Bijsal, laurel, teak for the retrieval of toxic heavy metal ions from dilute aqueous waste solution.

Magnetic nano-particle is an effective reliable method to remove heavy metal contaminants from wastewater through magnetic separation techniques. It has

1. High compactness.
2. High elimination performance
3. Low power input for magnetic operation.
4. Non-interference of biological system.
5. Successfully separate phosphate from sewage treatment plant and effluents.

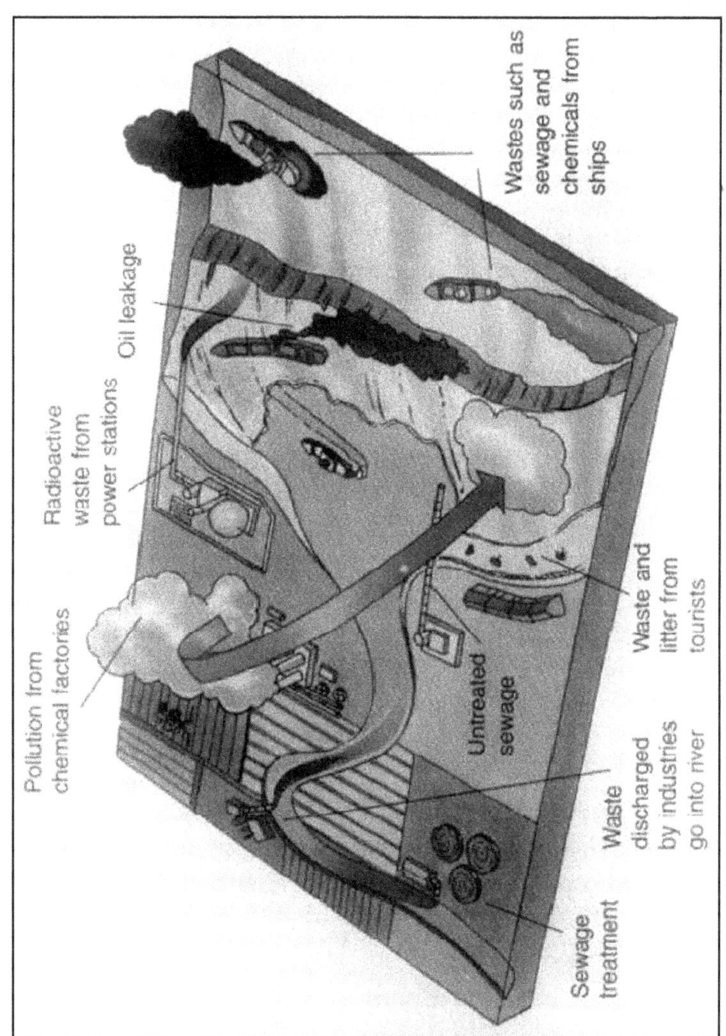

Figure 13.3: Different Sources of Pollution

Radioactive pollutant can be controlled by precipitation and coagulation technique and the high active solid wastes are dumped into the soil.

Plants like castor, common basil, cockroach plant, caraway, the plants belonging solanaceac family and alkaloids like nicotine etc. have effective insecticidal and fungicidal properties to control pests. Pests can be controlled by ecdysoncs and pheromone. Plant based botanical pesticides developed from plants like neem etc. The Scientist of CTCRI and Reekariyam at Thiruvanthapuram has recently developed tropica plants as pesticides.

Recycling of waste materials is a revolutionary step in the field of waste management.

1. The textile and chemical industries develop clothes which will be made from recycled plastic waste will come soon in global market.
2. Electricity is produced from the pile of garbage. The biodegradable and non biodegradable substances are separated from the garbage and mixed with water and then cultured with bacterial species to produce methane gas *i.e.* biogas and remains are used to produce manure. Bio-gas further produce electricity
3. Fuel is obtained from plastic wastes *i.e.* "*green fuel*" which has high octane rating.
4. Fly ash and slag from steel industries are utilized by cement industry.
5. The ashes from thermal power used as brick.

The water (Prevention and control pollution) Act 1974 and ammended in 1988 must be implemented seriously. Under section 24, there is prohibition on use of stream or well for disposal of polluting matter.

There are some suggestions to inhibit water pollution.

1. People must be educated and made aware of their environment. They must use environment- friendly chemicals.
2. Rain water should be stored to increase water-table.
3. Natural pesticides should be used as Agro-chemicals.

4. Plants should be set up to reuse the garbage and sewage waste to produce manure, electricity and biogas.
5 The old traditional technique should be used to remove turbidity from drinking water like crushed elaichi, plant ashes, earth from termite hill, paddy husk etc.
6. Solar appliances should be used such as solar cooker solar gyser, solar lamp etc.
7. The effluents of hospital, agriculture run off etc. should be led to the pits from where the water is slowly seep into the river.
8. Indian rivers and its tributaries are frequently flooded and thus dilute the concentration of water soluble pollutant and thus spread pollutants here and there to decrease the effect of pollution into the river-water.
9. Radio active substance can be separated from the waste by the method of coagulation or precipitation. Cow – dung has been found the best material to check radioactivity.

Acknowledgement

Author expresses her deep gratitude to Dr. Hashmat Ali, P. G. Deptt. of Chemistry, S. K. M. University, Dumka (Jharkhand) for his support. I also thank to my Principal, Dr. Ramanuj Prasad, D.A.V. School, B.S.E.B., Patna (Bihar) for encouragement.

References

1. Odum, Eugene P. *Ecology*, 2nd edn. Oxford and IBH Publishing Co. Pvt. Ltd.
2. Gupta, Kiran. *Encyclopedia of Knowledge*. Young Learner Publication, Y–68.
3. Gupta, Kiran. *Encyclopedia of Knowledge*. Young Learner Publication, Y–70.
4. Lee, J.D. (1996). *Concise Inorganic Chemistry*, 5th Edn. Blackwell Science Ltd., Publication.
5. *NCERT Textbook for Class XI, Chemistry Part II.*
6. *"The Times of India"*, Patna, March 16, 2011 (Wednesday). Page–1.

7. *"The Hindu"*, Kolkata, April 9, 2011 (Saturday) Editorial page–8.
8. *"The Times of India"*, Patna, April 12, 2011 (Tuesday) Page–1.
9. Kudesia, V.P. and Kudesia, Ritu. *Water Pollution*, Pragati Prakashan.
10. *Science Reporter*, Feb. 2011.

Chapter 14
Heavy Metal Toxicity

Awantika Kumari

Introduction

Heavy metals are the non- degradable metals.These metals are toxic and posses high density. Heavy metals occur in the earth's crust naturally. Some of the heavy metals are lead, cadmium, mercury, arsenic, chromium and thallium. High concentration of heavy metals causes poisoning. The improper disposal of these heavy metals leads to pollution. Heavy metal causes severe water pollution. Heavy metal poisoning is the accumulation of heavy metals, in toxic amounts, in the soft tissues of the body. Symptoms and physical findings associated with heavy metal poisoning vary according to the metal accumulated. Many of the heavy metals, such as zinc, copper, chromium, iron and manganese, are essential to body function in very small amounts. But, if these metals accumulate in the body in concentrations sufficient to cause poisoning, then serious damage may occur. The heavy metals most commonly associated with poisoning of humans are lead, mercury, arsenic and cadmium. Heavy metal poisoning may occur as a result of industrial exposure, air or water pollution, foods, medicines, improperly coated food containers, or the ingestion of lead-based paints.

Explanation

The heavy metals are

- ☆ Mercury
- ☆ Lead

- Cadmium
- Arsenic
- Chromium
- Thallium

Small amounts of heavy metals such as copper, zinc entered into our bodies are helpful in maintaining human metabolism. If the concentration gets increases, it will lead to poisoning effect. The common heavy metal pollution in drinking water contamination.

Sources

The main sources of heavy metal are weathering of rocks and volcanic eruption in which metals are released in water sources. Other sources are:

- Domestic wastewater and urban run-off
- Industrial wastewater
- Agricultural activities
- Mining activities

Domestic Wastewater and Urban Run-off

- Heavy metals are present in the domestic water. The metals like iron, manganese, nickel, zinc and chromium from detergents add to the wastewaters.
- This may be also due to corrosion of metal pipes which causes drinking water contamination.
- Run-off during the rainy season from the urban areas also contains heavy metals.

Industrial Wastewater

- Industries play the main source of producing heavy metals. The waste materials that contain heavy materials are disposed of to the environment causes water pollution.

Agricultural Activities

- Use of herbicides, pesticides and artificial fertilizers will increase the content of heavy metals in the soil. The soil erosion will take off the surface metals thus causes pollution in water bodies.

Mining Activities

- ✰ The waste rocks left at the mining places oxidized to release free metals to the environment causes heavy metal poisoning.

Environmental Effects

- ✰ Heavy metal pollution greatly infects aquatic organisms, small amount of lead leads to toxic effect in aquatic environment. This causes respiratory disorders in fishes. Thus the mortality of fishes increases with amount of lead.

Due to corrosion metal pipes, the drinking water gets contaminated with occurrence of heavy metals. Whenever the humans and animals consume that water, the lead present in tha t water will settle in their bones. They release very slowly from the human body.

Water Pollution Related with Heavy metal Poisoning

Mercury Poisoning: Treatment

Sign and Symptoms of Mercury Poisoning

Chronic exposure to metallic mercury vapor produces characteristic intension tremor. It can also produce *Mercurial erethism* (excitability, insomnia, timidity, memory loss, and delirium known as "*mad as a hatter*"). Decreased motor speed, visual scanning, and decreased verbal and visual memory, visuomotor (visual and motor) coordination are seen on neurobehavioral tests. Toxicity from elemental or inorganic mercury exposure begins when blood level is more than 3.6 µg/100 ml and urine levels more than 15µg/100 ml.

Organic mercury exposure is measured by mercury levels in blood in acute poisoning and mercury levels in hair in chronic poisoning.

If children are exposed to mercury in any form (organic, inorganic, vaporized or ingested) may develop *acrodynia* known as "*pink disease*"that include flushing, itching, swelling, irritability, hypertension, high pulse rate, excessive salivation, perspiration, weakness, morbilliform rashes, desquamation of palms and soles.

Treatment

Acute ingestion of mercury is treated by gastric lavage or by inducing vomiting (with gag reflex which is by touching the pharynx or by hypertonic saline or drugs that cause vomiting). Polythiol resin is given which binds to mercury in the gastrointestinal tract (GIT) and reduce absorption from GIT. *Chelating agent (bind metals into stable cyclic compounds with relatively low toxicity and enhances their excretion)* like dimercaprol (*British Anti Lewisite* or BAL), 24 mg/kg per day IM in divided doses, penicillamine or DMSA (succimer or dimercaptosuccinic acid) should be given. Chelating agents are given as several 5 days courses which are separated by few days of rest days. If renal failure develops, treat with hemodialysis or peritoneal dialysis.

Chronic inorganic mercury poisoning is treated with N-acetyl penicillamine.

Heavy Metal Poisoning: Mercury

Mercury is the only metal which is present in liquid form. The toxicity of low level organic mercury exposure is mainly manifested by neurobehavioral performance. Mercury is excreted in urine and feces and has a ½ life in blood of about 60 days. But, deposits in the kidney and brain may remain for many years. Elemental mercury (Hg°) is not absorbed well. But, it can volatilize into highly absorbable vapor. Inorganic mercury is absorbed through the gastro intestinal tract and skin. Organic mercury is well absorbed by ingestion and inhalation and it is a major source of mercury poisoning.

Sources of Mercury Poisoning

Metallic, mercuric mercury (Hg°, Hg^{+}, Hg^{2+}) and mercurous mercury exposures occur in some chemical, metal-processing, electrical equipment industries and automotive industries. Mercury is also present in thermometers, dental amalgams and batteries. Mercury can also be spread by waste incineration. Mercury present in environment is converted to organic mercury from inorganic mercury by bacteria. This organic mercury is than taken up by planktons, algae and fungi which are food for sea fishes like tuna, swordfish, and other *pelagic fish*. These sea foods when consumed by humans in large amount can lead to slow mercury poisoning.

Toxicity

Acute inhalation of mercury vapor can cause pneumonitis and pulmonary edema (water in lungs) which may cause death. It can also cause polyneuropathy and CNS (central nervous system) symptoms. Acute ingestion of inorganic mercury can cause gastroenteritis, nephritic syndrome, acute renal failure, hypertension, and cardiovascular collapse. Death usually occurs at a dose of 10–42 mg/kg. Acute ingestion of organic mercury causes gastroenteritis, arrhythmias (rhythm disturbance of heart beat), and lesions in the basal ganglia and gray matter. Chronic inhalation of mercury vapor causes CNS toxicity (mercurial erethism) lower exposures impair renal function, motor speed, memory, coordination.

High exposure of mercury during pregnancy can cause severe mental retardation due to derangement of fetal neuronal migration. Mild exposures of mercury during pregnancy (from fish like tuna, swordfish, and other *pelagic fish* consumption) are associated with reduced neurobehavioral performance in offspring.

Dimethylmercury is highly toxic (*supertoxic*) and is found in research labs. A few drops of exposure via skin absorption or inhaled vapor can cause severe cerebellar degeneration which lead to death.

Lead Poisoning: Treatment

Clinical Manifestations of Lead Poisoning

Abdominal pain, lethargy, anorexia, irritability, anemia, *Fanconi's syndrome*, pyuria (pus in urine), azotemia in children with blood lead level of more than 80μg/100ml. Epiphyseal plate *"lead lines"* can be seen on X-ray of long bones. Convulsions, coma, and death can occur if blood lead level is more than 120μg/100ml. CDC Atlanta, USA recommends screening of all children at the time of crawling age (about 6 months) source identification and intervention is begun if the BPb (blood lead level) is more than 10 μg/100ml. Neurodevelopmental delays are seen at BPb of 40–80 μg/100ml. Headaches, arthralgias (joint pain), myalgias, depression, impaired short-term memory, *loss of libido* are common symptoms of lead poisoning. Examination may reveal a *"lead line" at the gum-tooth border*, pallor, wrist drop.

Diagnosis

Diagnosis is mainly by history, clinical symptoms and blood lead levels. Laboratory tests may reveal a normocytic, normochromic anemia, an elevated blood protoporphyrin level, and motor delays on nerve conduction. In the U.S., regular testing of lead-exposed workers with removal if BPb is more than 40 μg/100ml is mandatory. *K-X-ray fluorescence* (*KXRF*) instruments have made it possible to measure bone lead levels which can diagnose a chronic lead poisoning even if it is at subclinical level.

Treatment

Source of the poisoning should be identified and corrected. Chelation is recommended with oral DMSA (succimer). Severe toxic cases should be hospitalized and IV (intravenous) or IM (intramuscular) chelation with *edentate calcium disodium* (*CaEDTA*) is administered. Dimercaprol is given to prevent worsening of encephalopathy. Correction of dietary deficiencies of iron, calcium, magnesium, and zinc lower lead absorption and also can improve the toxic condition. Vitamin C is a weak and natural chelating agent.

Chelation should be done or not in children with asymptomatic lead poisoning (blood lead level 20-40 μg/100ml) are not clear.

Heavy Metal Poisoning: Lead

U.S. Agency for Toxic Substances and Disease Registry lists toxic substances according to their prevalence and the severity of their toxicity and lead is in number one position. Recently the development of *K-X-ray fluorescence* (*KXRF*) instruments has made it possible to measure *bone lead levels* (which, in turn, reflect cumulative exposure over many years, as opposed to *blood lead levels*, which reflect recent exposure). High bone lead levels measured by KXRF are associated with increased risk of hypertension in both men and women. High maternal bone lead levels were found to cause lower birth weight, head circumference, birth length, and lower neurodevelopmental performance in children by age 2 years.

Sources of Lead Poisoning

Manufacturing of auto batteries, ceramics, fishing weights, lead crystal, demolition of lead-painted houses and bridges; stained glass making, soldering, environmental exposure to paint chips, plumbing, firing ranges (from bullet dust), food or water from lead pipes are the main sources of lead poisoning. Contaminated herbal remedies, candies and exposure to the combustion of leaded fuels also contribute to the lead poisoning.

Lead can be absorbed through ingestion or inhalation and organic lead (e.g., tetraethyl lead) is absorbed through skin. In blood lead is concentrated in RBCs. Distributed in soft tissue, with ½ life of about 30 days. 15 per cent of lead is sequestered in bone with ½ life of more than 20 years. Lead is excreted mainly in urine, but also appears in other fluids including *breast milk.*

Toxicity

Acute poisoning with *blood lead levels* (*BPb*) of more than 60–80 μg/100ml can cause impaired neurotransmission and neuronal cell death, which lead to central and peripheral nervous system effects. If blood lead level is more than 80 μg/100ml it can cause acute encephalopathy with convulsions, coma, and death.

Subclinical exposure of lead in children (BPb 25–60 μg/100ml) is associated with anemia, mental retardation, language deficit, motor function, balance, behavior, hearing, and school performance. Impairment of IQ can occur at even lower levels.

In adults, chronic subclinical exposures (BPb 40 μg/100ml) are associated with an increased risk of anemia, demyelinating peripheral neuropathy (mainly motor), and impairments of reaction time, hypertension, and ECG conduction delays. Chronic renal failure, diminished sperm counts and spontaneous abortions in females is seen.

Heavy Metal Poisoning: Cadmium

Cadmium poisoning can be a serious health problem from mining of cadmium. There was serious cadmium poisoning from contamination of food and water by mining effluents in Japan, in 1946 lead to outbreak of *"itai-itai"* (*"ouch-ouch"*) disease. The disease was so named because of cadmium-induced bone toxicity that led to painful bone fractures.

Sources of Arsenic Poisoning

Metal plating, battery, pigment, smelting, and plastics industries and incineration of these products are the main sources of cadmium poisoning. Tobacco smoking and consumption of food that concentrate cadmium like grains and cereals are also important source of cadmium poisoning.

Clinical Manifestations

Acute cadmium inhalation causes pneumonitis 4–24 hours after inhalation and acute ingestion causes gastroenteritis. Chronic exposure causes anosmia (loss of smell), yellowing of teeth, emphysema, microcytic hypochromic anemia that do not respond to iron therapy, proteinuria (protein in urine), calciuria (calcium crystals in urine), leading to chronic renal failure, osteomalacia, and fractures.

Symptoms of cadmium poisoning due to inhalation include chest pain, breathlessness, fever, pulmonary edema, nausea and high pulse rate. Symptoms due to ingestion are nausea, vomiting, cramps, and diarrhea.

Diagnosis

If poisoning is due to recent exposure, serum cadmium is about 5µg/dL. Urinary cadmium (10µg/g creatinine) and/or urinary $?_2$-microglobulin more than 750µg/g creatinine (but urinary $?_2$-microglobulin also increased in other renal diseases such as pyelonephritis, so it is not reliable).

Treatment

There is no effective and specific treatment forcadmium poisoning. Chelation is not useful and dimercaprol can aggravate renal toxicity. So the main management is further avoidance of exposure to cadmium and supportive therapy. Vitamin D is given for osteomalacia.

Heavy Metal Poisoning: Arsenic

Metals like arsenic pose a significant threat to health through occupational as well as environmental exposures. *U.S. Agency for Toxic Substances and Disease Registry* ranks arsenic third, according to,its prevalence and severity.

Sources of Arsenic Poisoning

Smelting and microelectronics industries, pesticides, fungicides, herbicides, contamination of deep-water wells, coal, incineration of these products and folk remedies are the main sources of arsenic poisoning. *Water pollution* is a source of arsenic poisoning.

Acute arsenic poisoning can cause necrosis of intestinal mucosa with resulting hemorrhagic gastroenteritis, fluid loss, hypotension, delayed cardiomyopathy, acute tubular necrosis and hemolysis. Chronic arsenic poisoning can cause diabetes, vasospasm (spasm of blood vessels), peripheral gangrene due to peripheral vascular insufficiency, peripheral neuropathy, and cancer of skin, liver, lungs, bladder and kidney.

Lethal Dose

The lethal dose of arsenic is 120–200 mg in adults and 2 mg/kg in children.

Clinical Manifestations

Nausea, vomiting, abdominal pain, diarrhea, coma, delirium and seizure can be seen. Typical garlic like odor is characteristic sign in arsenic poisoning. *Mees' lines* (transverse white striae of the fingernails), hyperkeratosis, hyperpigmentation, sensory and motor polyneuritis and distal weakness also seen.

Diagnosis

Radiopaque sign on abdominal X-ray; ECG shows QRS broadening, QT prolongation, ST depression, T-wave flattening; 24-h urinary arsenic 50 μg/day; (no seafood x 24 h); if recent exposure, serum arsenic 7 μg/100ml. High arsenic in hair or nails.

Treatment

In acute poisoning *ipecac* is given to induce vomiting. Gastric lavage (remove stomach content) and activated charcoal is given along with symptomatic treatment. *Dimercaprol* is given every 4 hourly at the dose of 3-5 mgs/kg intramuscularly for two days and every 6 hourly for 1 day, than twice a day for 10 days. Alternative to dimercaprol is *oralsuccimer*.

Heavy Metal Poisoning: Uncommon Types

Heavy Metal Poisoning by arsenic, cadmium, lead and mercury has been already discussed. Here we will discus about heavy metal poisoning by uncommon poisoning by heavy metals. Uncommon heavy metal poisoning includes poisoning by aluminum, chromium, cobalt, manganese, nickel, selenium, thallium, tin etc. The poisoning due to the above mentioned heavy metals is relatively rare and there is uncertainty regarding their potential toxicities.

Aluminum Poisoning

It can cause encephalopathy in patients with severe renal disease who are undergoing dialysis. High levels of aluminum are found in the cerebral cortex and hippocampus of patients with Alzheimer's disease. High levels of aluminum are also found in the drinking water and soil of areas with a high incidence of Alzheimer's disease. But it cannot be proved that aluminum is the causative factor or contributing factor in the development of Alzheimer's disease.

Chromium Poisoning

It is a corrosive. Workers of chromate and chrome pigment production industries have a greater risk of lung cancer due to chronic exposure to chromium. Hexavalent chromium is mainly responsible.

Cobalt Poisoning

Cobalt chloride was once used as fortifier of beer which led to outbreaks of fatal cardiomyopathy among heavy consumers. Now it is no more used and there are very rare incidences of cobalt poisoning due to it.

Manganese Poisoning

Chronic exposure to manganese can cause Parkinsonian syndrome. Parkinsonian syndrome is seen in persons like miners, dry-battery manufacturers, and arc welders. It is seen within 1–2 years of occupational exposure. Gait disorders, postural instability, tremor, expressionless face and psychiatric symptoms can be seen.

Nickel Poisoning

Nickel exposure can cause allergic reaction and chronic exposure by inhalation of nickel compounds with low aqueous solubility like nickel subsulfide and nickel oxide in occupational settings can cause is an increased risk of lung cancer.

Selenium Poisoning

Overexposure to selenium can cause local irritation of the respiratory system, gastrointestinal and eyes irritation, hepatitis, loss of hair, depigmentation, and peripheral nerve damage.

Thallium Poisoning

Thallium can be absorbed through ingestion, inhalation and also through skin. Thallium is used in insecticides, metal alloys, and fireworks. Severe poisoning occurs after a single ingested dose of more than 1g or more than 8 mg/kg. Nausea and vomiting, abdominal pain, and blood in vomit occur before confusion, psychosis and coma.

Treatment

The principle of treatment is same like other heavy metal poisoning. Chelating agents and symptomatic treatment should be given. Gastric lavage (removing stomach contents) can be done if poisoning is by ingestion. Being metal all are radio opaque and X-ray helps in diagnosis and extent of heavy metal in stomach.

In thallium poisoning *Prussian blue* prevents absorption and is given orally at 250 mg/kg in divided doses. Thallium poisoning may be less severe when activated charcoal is used to interrupt its *enterohepatic circulation* (liver and intestinal circulation of poison). Other measures include forced diuresis, treatment with potassium chloride to promote renal excretion of thallium, and peritoneal dialysis.

References

1. Aaseth, J., Jacobsen, D., Andersen, O. and Wickstron, E. (1995). Treatment of mercury and lead poisonings with dimercaptosuccinic acid (DMSA) and sodium dimmercapto-propanesulfonate (DMPS). *Analyst*, Mar, 120: 853.
2. AHA (2001). *Chelation Therapy. AHA Recommendation.* Dec 3. Dallas, TX: American Heart Association.
3. Anderton, R.M. (2001). *ADA Statement on Dental Amalgam.* May, Chicago, IL: American Dental Association.
4. Anon. (1993). Alzheimer's and aluminum: Canning the myth. *Food Insight*, Sept.–Oct. International Food Information Council Foundation, Washington, D.C.
5. Anon. (2001). *Management of the Poisoned/Overdosed Patient.* Continuing Education (No. 430–000–99–026–H01). U.S. Pharmacist, New York: Jobson.

6. Anuradha, B. and Varalakshmi, P. (1999). Protective role of DL-alpha-lipoic acid against mercury-induced neural lipid peroxidation. *Pharmacol. Res.*, Jan, 39(1): 67–80.

7. Bardin, J.A., Eisen, E.E., Wegman, D.H., Kriebel, D. Woskie, S.R. and Gore, R.J. (2000). Case–Control Studies of Liver, Gallbladder and Pancreatic Cancer and Metalworking Fluid Exposure in the Automobile Industry. Paper presented to the *128th Annual Meeting of the American Public Health Association*, November 14. Washington, D.C.: American Public Health Association.

8. Beers, M.H. and Berkow, M.D. (1999). *The Merck Manual of Diagnosis and Therapy*. Section 23. Chapter 307. Poisoning. Whitehouse Station, NJ: Merck and Co.

9. Brink, W. (2000). Lactoferrin: The bioactive peptide that fights disease. *Life Extension Magazine*, Oct, 6(10): 20–6. Ft. Lauderdale, FL: Life Extension Foundation.

10. Brown, D.J. (1998). Characterizing Risk at Metal Finishing Facilities. Report EPA/600/R–97/111 May 1998. Washington, D.C.: U.S. Environmental Protection Agency.

11. Cai, L., Koropatnick, J. and Cherian, M.G. (2001). Roles of vitamin C in radiation-induced DNA damage in presence and absence of copper. *Chem. Biol. Interact.* 2001 Jul 31, 137(1): 75–88.

12. Cha, C.W. (1987). A study on the effect of garlic to the heavy metal poisoning of rat. *J. Korean Med. Sci.*, Dec, 2(4): 213–24.

13. Chouchane, S. and Snow, E.T. (2001). *In vitro* effect of arsenical compounds on glutathione-related enzymes. *Chem. Res. Toxicol.*, May, 14(5): 517–22.

14. Clarkson, T.W. (1990). Mercury: An element of mystery. *N. Engl. J. Med.*, 323: 1137–1139.

15. Clayman, C.B. (Ed.) (1989). *The American Medical Association Encyclopedia of Medicine*. Random House, New York.

16. Cruz, T., Galvez, J. Ocete, M.A., Crespo, M.E., Sanchez de Medina, L.-H.F. and Zarzuelo, A. (1998). Oral administration of rutoside can ameliorate inflammatory bowel disease in rats. *Life Sci.*, 62(7): 687–695.

17. Daggett, E.A., Oberley, T.D., Nelson, S.A., Wright, L.S., Kornguth, S.E. and Siegel, F.L. (1998). Effects of lead on rat kidney and liver: GST expression and oxidative stress. *Toxicology*, Jul. 17, 128(3): 191–206.

18. De Flora, S., Izzotti, A., D'Agostini, F. and Balansky, R.M. (2001). Mechanisms of N-acetylcysteine in the prevention of DNA damage and cancer, with special reference to smoking-related end-points. *Carcinogenesis*, Jul, 22(7): 999–1013.

19. Deschner, E.E., Ruperto, J.F., Wong, G.Y. and Newmark, H.L. (1993). The effect of dietary quercetin and rutin on AOM-induced acute colonic epithelial abnormalities in mice fed a high-fat diet. *Nutr. Cancer*, 20(3): 199–204.

20. Dhir, H., Roy, A.K. and Sharma, A. (1993). Relative efficiency of *Phyllanthus emblica* fruit extract and ascorbic acid in modifying lead and aluminium-induced sister-chromatid exchanges in mouse bone marrow. *Environ. Mol. Mutagen*, 21(3): 229–236.

21. Dhir, H., Roy, A.K., Sharma, A. and Talukder, G. (1990). Modification of clastogenicity of lead and aluminium in mouse bone marrow cells by dietary ingestion of *Phyllanthus emblica* fruit extract. *Mutat. Res.*, Jul., 241(3): 305–12.

22. Dr. Joseph F. (2001). Smith Medical Library. *Heavy Metal Poisoning*, Nov. Wassau, WI: Medical Library/Thomson.

23. Dupler, D. (2001). Heavy metal poisoning. *Gale Encyclopedia of Alternative Medicine*. Farmington Hills, MI: Gale Group.

24. Esteves, A.C. and Felcman, J. (2000). Study of the effect of the administration of Cd(II), cysteine, methionine, and Cd(II) together with cysteine or methionine on the conversion of xanthine dehydrogenase into xanthine oxidase. *Biol. Trace Elem. Res.*, Jul., 76(1): 19–30.

25. Ewan, K. and Pamphlett, R. (1996). Increased inorganic mercury in spinal motor neurons following chelating agents. *Neurotoxicology*, 17(2): 343–349.

26. FDA. DMPS (1999). Washington, D.C.: Food and Drug Administration (http: //www.fda.gov/cder/fdama/pclist.txt).

27. Ferner, D.J. (2001). Toxicity, heavy metals. *eMed. J.* May 25, 2(5): 1.

28. Fournier, L., Thomas, G. and Garnier, R. *et al.* (1988). 2,3-Dimercapto-succinic acid treatment of heavy metal poisoning in humans. *Med. Toxicol. Adverse Drug Exp.*, 3: 499–504.

29. Galvez, J., Cruz, T., Crespo, E., Ocete, M.A., Lorente, M.D., Sanchez de Medina, F. and Zarzuelo, A. (1997). Rutoside as mucosal protective in acetic acid-induced rat colitis. *Planta Med.*, 63(5): 409–414.

30. Gebel, T. and Dunkelberg, H. (1996). Influence of chewing gum consumption and dental contact of amalgam fillings to different metal restorations on urine mercury content. *Zentralbl. Hyg. Umweltmed*, Nov, 199(1): 69–75 (in German).

31. Ghio, A.J., Kennedy, T.P., Crissman, K.M., Richards, J.H. and Hatch, G.E. (1998). Depletion of iron and ascorbate in rodents diminishes lung injury after silica. *Exp. Lung Res.*, Mar–Apr., 24(2): 219–232.

32. Girodon, F., Galan, P. Monget, A.L., Boutron-Ruault, M.C., Brunet-Lecomte, P., Preziosi, P., Arnaud, J., Manuguerra, J.C. and Herchberg, S. (1999). Impact of trace elements and vitamin supplementation on immunity and infections in institutionalized elderly patients: A randomized controlled trial. MIN.VIT.AOX. geriatric network. *Arch. Intern. Med.*, Apr 12, 159(7): 748–54.

33. Glanze, W.D. (1996). *Mosby Medical Encyclopedia*, Revised Edition. St. Louis MO: C.V. Mosby.

34. Goering, P.L., Aposhian, H.V., Mass, M.J., Cebrian, M., Beck, B.D. and Waalkes, M.P. (1999). The enigma of arsenic carcinogenesis: Role of metabolism. *Toxicol. Sci.*, May, 49(1): 5–14.

35. Gonzalez-Correa, J.A., de la Cruz, J.P., Gordillo, J., Urena, I., Redondo, L. and Sanchez de la Cuesta, F. (2002). Effects of silymarin MZ–80 on hepatic oxidative stress in rats with biliary obstruction. *Pharmacology*, Jan, 64(1): 18–27.

36. Goyer, R.A. (1996). Toxic effects of metals: Mercury. *Casarett and Doull's Toxicology: The Basic Science of Poisons*, 5th edn. McGraw–Hill, New York.

37. Gubrelay, U., Mathur, R., Kannan, G.M. and Flora, S.J. (2001). Role of S-adenosyl-L-methionine in potentiating cadmium

mobilization by diethylenetriamine penta acetic acid in mice. *Cytobios*, 104(406): 99–105.

38. Gurer, H., Ozgunes, H., Oztezcan, S. and Ercal, N. (1999). Antioxidant role of alpha-lipoic acid in lead toxicity. *Free Radic. Biol. Med.*, Jul, 27(1–2): 75–81.
39. Horikoshi, T., Nakajima, A. and Sakaguchi, T. (1979). Uptake of uranium by various cell fractions of *Chlorella regularis*. *Radioisotopes*, Aug, 28(8): 485–488.
40. Huang, K.-C. (1993). *The Pharmacology of Chinese Herbs*. CRC Press, Boca Raton, Fl.
41. Ichimura, S. (1973). Report. General meeting of the Pharmaceutical Society of Japan, Hokuriku Branch, Toyoma City, Japan, October 27.
42. International Occupational Safety and Health Information Centre. Metals 1999 Sep. Geneva: International Labour Organization.
43. Isacsson, G., Barregard, L., Selden, A., Bodin, L. Impact of nocturnal bruxism on mercury uptake from dental amalgams. *Eur. J. Oral Sci.* 1997 Jun, 105(3): 251–7.
44. James, D. (2001). *General Methods for Treating Poisoning*. Alberta, Canada: University of Alberta.
45. Klein-Schwartz, W. and Oderda, G.M. (2000). Clinical toxicology. *Textbook of Therapeutics: Drug and Disease Management*, 7th Edn. Baltimore, MD: Williams and Wilkins, p. 51.
46. Kostyuk, V.A. and Potapovich, A.I. (1998). Antiradical and chelating effects in flavoinoid protection against silica-induced cell injury. *Arch. Biochem. Biophys.*, Jul 1, 355(1): 43–48.
47. Kostyuk, V.A., Potapovich, A.I., Speransky, S.D. and Maslova, G.T. (1996). Protective effect of nautral flavonoids on rat peritoneal macrophages injury caused by asbestos fibers. *Free Radic. Biol. Med.*, 21(4): 487–493.
48. Leung, F.Y. (1998). Trace elements that act as antioxidants in parenteral micronutrition. *Can. J. Nutr. Biochem.*, 9(6): 304–307.
49. Lide, D. (1992). *CRC Handbook of Chemistry and Physics*, 73rd Edn. CRC Press, Boca Raton, FL.

50. Lorico, A., Bertola, A., Baum, C., Fodstad, O. and Rappa, G. (2002). Role of the multidrug resistance protein 1 in protection from heavy metal oxyanions: investigations *in vitro* and in MRP1–deficient mice. *Biochem. Biophys. Res. Commun.*, Mar 1, 291(3): 617–622.

51. Lupton, G., Kao, G. and Johnson, F. *et al.* (1985). Cutaneous mercury granuloma: A clinicopathologic study and review of the literature. *J. Am. Acad. Dermatol.*, 12: 296–303.

52. Maiti, S. and Chatterjee, A.K. (2001). Effects on levels of glutathione and some related enzymes in tissues after an acute arsenic exposure in rats and their relationship to dietary protein deficiency. *Arch. Toxicol.*, Nov, 75(9): 531–537.

53. Marcus, S. (2001). Toxicity, lead. *eMed. J.*, Jun 4, 2(6): 7.

54. Medical Management Guidelines (MMGs). Managing Hazardous Material Incidents, Volume III 2001. Atlanta, GA: Agency for Toxic Substances and Disease Registry (http: // www.atsdr.cdc.gov).

55. Micromedex. B.A.L.™ (1999). Greenwood Village, CO: Thomson.

56. Milchak, L.M. and Douglas Bricker, J. (2002). The effects of glutathione and vitamin E on iron toxicity in isolated rat hepatocytes. *Toxicol. Lett.*, Feb 7, 126(3): 169–77.

57. Muller, L. (1989). Protective effects of DL-alpha-lipoic acid on cadmium-induced deterioration of rat hepatocytes. *Toxicology* Oct 2, 58(2): 175–185.

58. Muller, L. and Menzel, H. (1990). Studies on the efficacy of lipoate and dihydrolipoate in the alteration of cadmium 2^+ toxicity in isolated hepatocytes. *Biochim. Biophys. Acta* 1990 May 22, 1052(3): 386–91.

59. Muller, U. and Krieglstein, J. (1995). Prolonged pretreatment with alpha-lipoic acid protects cultured neurons against hypoxic, glutamate-, or iron-induced injury. *J. Cereb. Blood Flow Metab.* Jul, 15(4): 624–30.

60. National Medical Library. Poisoning first aid. Medical Encyclopedia 2001. Bethesda, MD: National Institutes of Health.

61. O'Brien, J. (2001). Mercury amalgam toxicity. *Life Extension Magazine 2001* May. 7(5): 43–51. Ft. Lauderdale, FL: Life Extension Foundation.

62. Omura, Y. and Beckman, S.L. (1995). Role of mercury (Hg) in resistant infections and effective treatment of Chlamydia trachomatis and Herpes family viral infections (and potential treatment for cancer) by removing localized Hg deposits with Chinese parsley and delivering effective antibiotics using various drug uptake enhancement methods. *Acupunct. Electrother. Res.*, 20(3–4): 195–229.

63. Omura, Y., Shimotsuura, Y., Fukuoka, A., Fukuoka, H. and Nomoto, T. (1996). Significant mercury deposits in internal organs following the removal of dental amalgam, and development of pre-cancer on the gingiva and the sides of the tongue and their represented organs as a result of inadvertent exposure to strong curing light (used to solidify synthetic dental filling material) and effective treatment: A clinical case report, along with organ representation areas for each tooth. *Acupunct. Electrother. Res.*, 1(2): 133–160.

64. Pakdaman, A. (1998). Symptomatic treatment of brain tumor patients with sodium selenite, oxygen, and other supportive measures. *Biol. Trace Elem. Res.*, Apr–May, 62(1–2): 1–6.

65. Paredes, S.R., Kozicki, P.A. and Batlle, A.M. (1985). S-adenosyl-L-methionine a counter to lead intoxication? *Comp. Biochem. Physiol.*, B, 82(4): 751–757.

66. Perez Guerrero, C., Martin, J.J. and Marhuenda, E. (1994). Prevention by rutin of gastric lesions induced by ethanol in rats: Role of endogenous prostaglandins. *Gen. Pharmacol.*, May, 25(3): 575–80.

67. Pietrangelo, A., Borella, F., Casalgrandi, G., Montosi, G., Ceccarelli, D., Gallesi, D., Giovannini, F., Gasparetto, A. and Masini, A. (1995). Antioxidant activity of silybin in vivo during long-term iron overload in rats. *Gastroenterology*, Dec, 109(6): 1941–9.

68. Porter, J.M., Ivatury, R.R., Azimuddin, K. and Swami, R. (1999). Antioxidant therapy in the prevention of organ dysfunction syndrome and infectious complications after trauma: early

results of a prospective randomized study. *Am. Surg.*, Sep, 65(9): 902.

69. Prater, G. (1999). MSM: The multi-purpose compound. *Life Extension Magazine* Sep, 5(9): 71–2. Ft. Lauderdale, FL: Life Extension Foundation.

70. Pryor, W.A. (2000). Vitamin E and heart disease: basic science to clinical intervention trials. *Free Radic. Biol. Med.*, Jan 1, 28(1): 141–64.

71. Quig, D. (1998). Cysteine metabolism and metal toxicity. *Altern. Med. Rev.* Aug, 3(4): 262–70.

72. Ringwood, A.H. and Conners, D.E. (2000). The effects of glutathione depletion on reproductive success in oysters, Crassostrea virginica. *Mar. Environ. Res.*, Jul–Dec, 50(1–5): 207–11.

73. Roberts, J.R. (1999). Metal Toxicity in Children. Training Manual on Pediatric Environmental Health: Putting it into Practice, Dec 10. Emeryville, CA: Children's Environmental Health Network.

74. Sallsten, G., Thoren, J., Barregard, L., Schutz, A. and Skarping, G. (1996). Long-term use of nicotine chewing gum and mercury exposure from dental amalgam fillings. *J. Dent. Res.*, Jan, 75(1): 594–8.

75. Saxe, S.R., Wekstein, M.W., Kryscio, R.J., Henry, R.G., Cornett, C.R., Snowdon, D.A., Grant, F.T., Schmitt, F.A., Donegan, S.J., Wekstein, D.R., Ehmann, W.D. and Markesbery, W.R. (1999). Alzheimer's disease, dental amalgam and mercury. *J. Am. Dent. Assoc.*, Feb, 130(2): 191–9.

76. Schumacher, K. (1999). Effect of selenium on the side effect profile of adjuvant chemotherapy-radiotherapy in patients with breast carcinoma. Design for a clinical study. *Med. Klin.*, Oct 15, 94(Suppl. 3): 45–8 (in German).

77. Shaikh, Z.A., Northup, J.B. and Vestergaard, P. (1999a). Dependence of cadmium-metallothionein nephrotoxicity on glutathione. *J. Toxicol. Environ. Health* A, Jun 11, 57(3): 211–22.

78. Shaikh, Z.A. and Tang, W. (1999b). Protection against chronic cadmium toxicity by glycine. *Toxicology*, Feb 15, 132(2–3): 139–46.

79. Shukla, G.S., Srivastava, R.S. and Chandra, S.V. (1988). Glutathione status and cadmium neurotoxicity: Studies in discrete brain regions of growing rats. *Fundam. Appl. Toxicol.*, Aug, 11(2): 229–235.

80. Sidhu, M., Sharma, M., Bhatia, M., Awasthi, Y.C. and Nath, R. (1993). Effect of chronic cadmium exposure on glutathione S–transferase and glutathione peroxidase activities in rhesus monkey: the role of selenium. *Toxicology*, Oct. 25, 83(1–3): 203–13.

81. Skottova, N., Krecman, V. and Simanek, V. (1999). Activities of silymarin and its flavonolignans upon low density lipoprotein oxidizability *in vitro*. *Phytother. Res.*, Sep, 13(6): 535–7.

82. Smith, S.R., Jaffe, D.M. and Skinner, M.A. (1997). Case report of metallic mercury injury. *Pediatr. Emer. Care*, 13: 114–116.

83. Smith-Barbaro, P., Hanson, D. and Reddy, B.S. (1981). Carcinogen binding to various types of dietary fiber. *J. Natl. Cancer Inst.*, Aug, 67(2): 495–7.

84. Sonnebichler, J., Goldberg, M., Hane, L., Madubunyi, I., Vogl, S. and Zetl, I. (1986). Stimulatory effect of silibinin on the DNA synthesis in partially hepatectomized rat livers: Non-response in hepatoma and other malign cell lines. *Biochem. Pharmacol.*, Feb 1, 35(3): 538–541.

85. Sonnebichler, J., Scalera, F., Sonnenbichler, I. and Weyhenmeyer, R. (1999). Stimulatory effects of silibinin and silicristin from the milk thistle Silybum marianum on kidney cells. *J. Pharmacol. Exp. Ther.*, Sep, 290(3): 1375–1383.

86. Stella, V. and Postaire, E. (1995). Evaluation of the antiradical protector effect of multifermented milk serum with reiterated dosage in rats. *C. R. Seances Soc. Biol. Fil.*, 189(6): 1191–7 (in French).

87. Tager, M., Dietzmann, J., Thiel, U., Hinrich Neumann, K. and Ansorge, S. (2001). Restoration of the cellular thiol status of peritoneal macrophages from CAPD patients by the flavonoids silibinin and silymarin. *Free Radic. Res.*, Feb, 34(2): 137–151.

88. Tandon, S.K., Singh, S. and Dhawan, M. (1992). Preventive effect of vitamin E in cadmium intoxication. *Biomed. Environ. Sci.*, Mar, 5(1): 39–45.

89. Tang, W., Sadovic, S. and Shaikh, Z.A. (1998). Nephrotoxicity of cadmium-metallothionein: protection by zinc and role of glutathione. *Toxicol. Appl. Pharmacol.*, Aug, 151(2): 276–282.

90. Tjalkens, R.B., Valerio, L.G., Jr., Awasthi, Y.C. and Petersen, D.R. (1998). Association of glutathione S–transferase isozyme–specific induction and lipid peroxidation in two inbred strains of mice subjected to chronic dietary iron overload. *Toxicol. Appl. Pharmacol.*, Jul, 151(1): 174–81.

91. ToxFAQs™ for Aluminum. CAS 7429–90–5. 1999 Jun. Atlanta, GA: Agency for Toxic Substances and Disease Registry.

92. ToxFAQs™ for Arsenic. CAS 7440–38–2. 2001 Jul. Atlanta, GA: Agency for Toxic Substances and Disease Registry.

Chapter 15

Water Pollution: A Threat to Humanity

Biswajit Mitra and Hashmat Ali

ABSTRACT

Water is the most valuable resource from the point of view of survival of almost all organisms on the earth. Its pollution, as a necessary corollary, draws our attention. Due to various anthropogenic activities the water bodies are getting intoxicated in varying extent. Industrialization, commercialization of agriculture, capitalistic mode of resource and land utilization has all increased the spread and pace of pollutant accumulation in water. Installation of industries generating hazardous waste material in the Developing Countries by the First World is intensifying the pollution problem of the under-developed nations. Juxtaposed to the pollution of surface water, groundwater is also getting contaminated day by day. While the aquatic community is being compelled to live in the polluted environment and perish gradually, so is the condition of the poverty stricken people of the Third World who have no other alternative but to live with the silent killers. Surface water purification methods, though available in numbers, are mostly costly, especially to make it potable. In developing countries, therefore, providing pure and safe drinking water in adequate quantity for all is indeed a challenging job. Cost effective and indigenous technologies may be emphasized upon in this regard. Developing consciousness and educating people, eradicating poverty and controlling population is of utmost importance to combat the problem. If the social autotrophs die can others be far behind?

Introduction

Water is also known as life as it is the resource of utmost necessity for the sustenance of life on the earth. Apart from drinking it is used for our domestic, agricultural, industrial, commercial and all other activities. However, water which we can easily access and easily use is limited.

The total amount of water in the world is estimated to be 1400 million cu. km. But 97 per cent of it is saline. Remaining 3 per cent is fresh water which is locked up in relatively inaccessible icecaps and glaciers. That leaves a mere 1 per cent or 14 million cu. km. Again half of this groundwater and most of it lies too far underground. About 200,000 cu. km. can be found in rivers and lakes and 14000 cu. km. in the atmosphere (Rajagopalan, 2005). So what ever little amount of easily available fresh water we have it is in the surface waterbodies like lakes, rivers, ponds.

According to the World Health organization (WHO) the minimum water requirement for domestic use is 50 litre/capita/day (though 100 l to 200 l is often recommended). Adding the needs of agriculture, industry, and the energy sector, the recommended minimum annual per capita requirement is about 1700 cu. m. In India, the per capita water availability is decreasing sharply. From 5177 cu. m. in 1951 it has dropped to 1820 cu. m. in 2001. We are near the threshold value of 1700 below which we will enter the list of countries facing periodic water stress.

Resource being limited and demand being high it is wise to keep the resource in useable condition, use it in a judicious and sustainable manner. However, contrary to what is desirable we are disposing sewage, agricultural and industrial wastewater in the waterbodies and thus contaminating them. Very often solid wastes are disposed there.

Pollution of Water

Water pollution is the contamination of water bodies (*e.g.* lakes, rivers, oceans and groundwater). Water pollution occurs when pollutants are discharged directly or indirectly into water bodies without adequate treatment to remove harmful compounds. Water is typically referred to as polluted when it is impaired by anthropogenic contaminants and either does not support a human use, such as drinking water, and/or undergoes a marked shift in its ability to support its constituent biotic communities, such as fish.

The rivers of India, as found by Bhradwaj (2005), are highly polluted. Some of his findings are shown in the following table:

Table 15.1: Water Quality in Indian Rivers–2002

River	*pH*	*DO (mg/l)*	*BOD (mg/l)*	*Faecal Coliform (MNP/100ml)*
Ganga	6.4-9.0	27-115	0.5-16.8	20-11*10^5
Yamuna	6.7-9.8	0.1-227	1.0-36	11-17.2*10^5
Sabarmati	2.9-8.6	0.6-7.9	0.8-475	28-28*10^5
Krishna	6.8-9.5	2.9-10.9	0.2-10.0	3-10000
Cauvery	2.0-9.2	0.1-12.6	0.1-26.6	2-280000
Godavari	7.0-9.0	3.1-10.9	0.5-78	2-3640
Subarnarekha	6.5-8.0	5.2-8.5	0.2-12.0	70-540
Brahmaputra	6.5-9.0	1.1-10.5	0.1-3.9	300-24000

Similar picture of water pollution may be seen almost all over the globe. Being the leading worldwide cause of deaths and diseases water pollution is a menace to the humanity, It accounts for the deaths of more than 14,000 people daily all over the globe (West,2006) most of which is in the developing countries. In India alone, for example, about 1,000 children die of diarrhea every day (The Economist, 2008). Developing countries of the Third World are faced with problem of rapid growth of population. To feed its large population and to go for economic development it is necessary for the developing countries increase agricultural production and industrialization. at the same time pollution to be kept in control to safeguard its citizens. But the countries are not only financially weak but the economic disparity of their citizens is very high. Hence, water pollution problem is more intricate and of more concern in the Third World countries with high population, paucity of resource, weak economy, dearth of modern technological know-how, lack of health and environmental consciousness, illiteracy and abject poverty.

Pollutants and their Effects

Major pollutants responsible for surface water pollution and their effect are as follows:

Pesticides

Pesticides that are used widely in the farm fields run off into local streams and rivers or drain down into groundwater, contaminating the fresh water. It jeopardizes the aquatic ecosystem and biodiversity there. It also has telling effect on human health. Pesticides from air and water accumulate in animal tissues, humus, soil in concentrations thousands of time higher than it is found in water through food chain. Pesticides like DDT, use of which is banned in the Developed nations but still widely used in the poor under-developed countries, is fatal for fish, crustaceans and shell-fish. Some organisms are extra-ordinarily sensitive even to very low concentration of DDT. For example, one part DDT per trillion seems to cause death in brine shrimps. The foremost danger of changing of behavior or metabolic activities of the body arises from DDT (Manivasakam,1984). It causes depression of photosynthesis in planktons. It makes fish unable to tolerate low temperatures. It causes birds to lay thin shelled eggs. Even in the raptorial hawks and falcons it cause death and birth failure. Developed countries spend billions of dollars to treat contaminated water. In the mid-western United States, a region that is highly dependent on groundwater, water utilities spend $400 million each year to treat water for just one chemical–the pesticide Atrazine.

Fertilizers

Another major cause of pollution, including sewage, manure, and chemical fertilizers, contain "nutrients" such as nitrates and phosphates. In excess levels, nutrients over-stimulate the growth of aquatic plants and algae. Excessive growth of these types of organisms turns an oligotrophic waterbody to eutrophic and hyper-eutrophic within a short span of time. When the organisms die, they use up dissolved oxygen as they decompose, causing oxygen-poor waters that support only diminished amounts of marine life. Such areas are commonly called dead zones. In an agricultural area nutrient pollution is of considerable problem and is responsible for the death of many large and small waterbodies.

Oil

Oil spills like the Exxon Valdez spill off the coast of Alaska or the more recent Prestige spill off the coast of Spain get lots of news coverage, and indeed they do cause major water pollution and problems for local wildlife, fishermen, and coastal businesses. But the problem of oil polluting water goes far beyond catastrophic oil spills. Land-based petroleum pollution is carried into waterways by rainwater runoff. This includes drips of oil, fuel, and fluid from cars and trucks; dribbles of gasoline spilled onto the ground at the filling station; and drips from industrial machinery. These sources and more combine to provide a continual feed of petroleum pollution to all of the world's waters, imparting an amount of oil to the oceans every year that is more than 5 times greater than the Valdez spill.

Mining

Mining causes water pollution in a number of ways. The mining process exposes heavy metals and sulfur compounds that were previously locked away in the earth. Rainwater leaches these compounds out of the exposed earth, resulting in acid mine drainage (AMD) and heavy metal pollution that can continue long after the mining operations have ceased. AMD contaminate drinking water and disrupt growth and reproduction of aquatic life. It also causes death of fish.

In the case of gold mining, cyanide is intentionally poured on piles of mined rock (a leach heap) to chemically extract the gold from the ore. Some of the cyanide ultimately finds its way into nearby water.

Sediment- Due to deforestation and barring ground eroded soil run off into nearby streams, rivers, and lakes. Increased amount of sediment influx into nearby waterbodies seriously affects fish and other aquatic life. The dissolved sediments reduce the amount of sunlight available to aquatic plants and retard their growth.

Industrial Wastes

Industrial and manufacturing activity has serious health effects in humans (Jorgenson 2004; Jorgenson and Burns 2004). Almost all waterbodies in the world have some level of pollution from chemicals and industrial waste. Different industrial waste products are often drained into the nearby rivers, seas or other waterbodies. Apart from different acids, alkalis, salts, chemicals, some industries often discharge effluents containing phenol, mercury, copper, lead, chromium, cadmium, zinc etc., which are very toxic to human beings and aquatic life. The toxic chemicals found in water affect people through the process of "bioaccumulation" i.e, the building up of toxins in the fatty tissue of mammals. The long term effects of bioaccumulation in adults include cancer, blood disorders, immunity suppression, and spontaneous abortions. Mercury affects our central nervous system. The inorganic mercury discharged by the industries is converted into methyl mercury by certain anaerobic bacteria. Mercury that can concentrate in the food chain are very harmful for man. The Minamata disease may be remembered here. In the year 1960 in Japan mercury poisoning caused irreversible damage to brain cell of 116 persons. In the subsequent years many deaths were reported there from due to mercury pollution. Cadmium which enters through food chain concentrates in liver, kidney and thyroid cause vomiting and diarrhea. Other metallic pollutants also enter our body through food chain and cause severe damage to our body.

Domestic Products and Cosmetics

Whenever we use personal-care products and household cleaning products–whether they be laundry detergent, bleach, or fabric softener; window cleaner, dusting spray, or stain remover; hair dye, shampoo, conditioner, or Rogaine; cologne or perfume; toothpaste or mouthwash; antibacterial soap or hand lotion–we should realize that almost all of it goes down the drain when we do laundry, wash our hands, brush our teeth, bathe, or do any of the other myriad things that incidentally use household water. Similarly, when we take medications, we eventually excrete the drugs in altered or unaltered form, sending the compounds into the waterways. Studies have shown that up to 90 per cent of your original prescription passes out of you unaltered. Animal farming operations that use growth hormones and antibiotics also send large quantities of these chemicals into our waters.

Unfortunately, most wastewater treatment facilities are not equipped to filter out personal care products, household products, and pharmaceuticals, and a large portion of the chemicals passes right into the local waterway that accepts the treatment plant's supposedly clean effluent.

Study of the effects of these chemicals getting into the water is just beginning, but examples of problems are now popping up regularly.

Sewage Water

In developing countries, an estimated 90 per cent of wastewater is discharged directly into rivers and streams without treatment. Even in modern countries, untreated sewage, poorly treated sewage, or overflow from under-capacity sewage treatment facilities can send disease-bearing water into rivers and oceans. Leaking septic tanks cause groundwater contamination.

Main chemicals in groundwater that cause health problems are arsenic and fluoride (the greatest problems of all inorganic constituents); nitrate (mainly from sanitation or agriculture) and metals which can be mobilized when the pH is extreme.

Fluoride

Impacts from long-term consumption of fluoride-bearing water are summarized below.

Table 15.2

Consumption (mg/l)	*Health Effect*
<0.5	Dental caries
0.5-1.5	Promotes dental health
1.5-4.0	Dental fluorosis
>4.0	Dental and skeletal fluorosis
>10.0	Skeletal fluorosis

Floride in water derives mainly from dissolution of natural minerals in the rocks and soils through which it passes. The most common fluorine-bearing minerals are fluorite, apatite and micas, and fluoride problems tend to occur where these elements are most abundant in the host rocks. Groundwater from crystalline rocks, especially granites, is particularly susceptible to fluoride build-up because they often contain abundant fluoride-bearing minerals. Reaction times with aquifer minerals are important. High fluoride concentration can build up in groundwaters that have long residence time in host aquifers. Deeper groundwaters from boreholes are likely to contain high concentrations of fluoride.

High-fluoride groundwaters are found in many parts of the developing world, and many million of people rely on groundwater with concentrations above the WHO guideline value. The worst affected areas are arid parts of northern China, India, Sri Lanka, Ghana, Ivory Coast, Senegal, Algeria, Kenya, Uganda, Tanzania, Ethiopia. In India alone, endemic fluorosis is thought to affect around 60 million people and is a major problem in 17 out of the country's 22 states, esp., Rajasthan, Andhra Pradesh, Tamil Nadu, Gujarat and Uttar Pradesh.

Arsenic

Arsenic has long been recognized as a toxin and carcinogenic. long-term ingestion of high concentrations from drinking-water can give rise to a number of health problems, particularly skin disorders such as pigmentation change and keratosis (warty noudels, usually on the hands and feet). Additional symptoms include more serious dermatological problems (eg., skin cancer and Bowen's disease); cardiovascular problems and Raynaud's syndrome; blackfoot disease and gangrene; neurological, respiratory,renal and hepatic diseases as well as diabetes mellitus. Internal cancers, particularly of the lung, bladder,liver,prostate and kidney have also been linked with arsenic in drinking water (Smith et.al. 1998) it can take years for arsenic related health problems to become apparent, which help to explain why many of the problems in developing countries have only recently emerged despite prolonged groundwater use.

Arsenic occurs naturally in a number of geological environments. It occurs in sulphide minerals precipitated from hydrothermal fluids, in pyrite accumulated sedimentary environments.

Groundwaters are generally more vulnerable to accumulation of high arsenic accumulation than surface water due to increased contact between rocks and water. Arsenic problems may occur in areas of mining of coal and metals associated with sulphide minerals. Under normal pH conditions, arsenic is strongly adsorbed onto sediments and soils, particularly iron oxides, as well as aluminium and manganese oxides and clay.

High concentrations of of arsenic in groundwater are, therefore, mainly found where adsorption is naturally inhibited under either two aquifer conditions:

Strongly reducing (no oxygen, anaerobic) groundwaters where arsenite dominates and hence adsorption to iron oxides is less favourable. Iron oxides themselves may also dissolve in such conditions, which may release further arsenic.

Oxidizing (aerobic) aquifers with high pH (>8), typically restricted to arid or semi-arid environments. Such groundwaters commonly also have high concentrations of other potentially toxic elements such as fluoride, boron, uranium, vanadium, nitrate and selenium.

In the south-eastern districts of West Bengal in India, the groundwater occurring mainly within the shallow zone (20-60m bgl) is characterized by high arsenic (>0.5 to 1 or above mg/l) and the principal source of arsenic is the arsenic sulphides minerals deposited alongwith clay, peat, with iron in the reducing environment. The lowering of groundwater at rapid rate during summer season causes aeration of aquifer oxidized the arsenic sulphides, makes it water soluble. It percolates from the subsoil into water table during monsoon.

Why Pollution?

In the above discussion we have discussed about different major sources of water pollution. Chemical wastes from different industries and agricultural sector are not only hazardous but sometimes cause disaster for aquatic and terrestrial population. Soil erosion and sedimentation also deteriorate the quality of water as it increases the turbidity and consequently affect the ecosystem. Changes in the quality of water are not new. Rather, it is a process, natural and anthropogenic, which is on since times immemorial. Pollution, literally speaking, is rather new and outcome of capitalistic mode of production. Profit maximization motive of the entrepreneurs resulted in rapacious use of resources, introduction of various exogenous elements in agricultural ecosystem to increase productivity, changes in landuse without paying heed to the tenacity of the ecosystem. Establishment of various industries at population sites and discharge of their untreated waste in the nearby waterbodies is also an outcome of the profit maximization motive.

Poverty and illiteracy which too are products of capitalistic system, are also responsible for water pollution. Existence of wealthy and poor under-developed nations at the same time has encouraged unequal exchange rate (Shandra *et al.*, 2009). Wealthy nations are advantageously situated within the global economy and are more likely to secure favorable terms of trade (Hornborg 2003; Bunker 1984). This advantage facilitates disproportionate access of wealthy nations to natural resources and ecological sink capacities of poor nations (Rice 2007; Hornborg 2001). Put differently, wealthy nations are able to shift many of the negative environmental externalities associated with their natural resource demands onto poor nations (Jorgenson 2006b; Jorgenson and Rice 2005; Bunker 1984). The unequal exchange can be demonstrated with case study evidence. In 1986, Sandoz, an agribusiness and pharmaceutical corporation, was responsible for the worst river spill in history in which 30 tons of extremely hazardous organophosphates, disulfoton and parathion, spilled into the Rhine in Switzerland (Karliner 1997). The spill killed fish, wildlife, and plants for hundreds of miles along the river. Karliner (1997, 129) writes, "Sandoz responded by cleaning up its operations and moving 60 per cent of its organophosphate production to Brazil. After another ton of disulfoton ended up in the Rhine in 1989, Sandoz moved its remaining pesticide production facilities to India." Today, disulfoton and parathion are exported largely to the United States for use on crops, gardens, and potted plants. Meanwhile, the chemicals have turned up in rivers and streams near the production facilities in India and Brazil (Karliner 1997).

As we have seen, most of the rivers in India and other developing countries are carrying enormous amount of pollutants. The Third World countries which remained long under indigenous landlords and colonial rulers are in such a socio-economic condition that they are bound to accept foreign technologies with all their ill-effects and terms and conditions as lay down by the developed nations.

Not only the industries run by the foreign companies and other entrepreneurs bit also industries ruin by the government of the countries concern are polluting water. Several examples can be cited to exemplify that it is the poverty which compel the poor to live and bear with water pollution. Their ignorance about the toxic chemicals and other pollutants, which cause slow poisoning, do not resist them from using polluted water. However, sometimes poverty compels them to use polluted water knowing. In some villages of the arsenic affected districts of West Bengal people often have other option but to drink arsenic affected water.

In the developing countries, in the vicious cycle of poverty, increase of population is a well known fact. Pressure of population is responsible for overuse of resources which in a way is responsible for water pollution. To feed the vast population of our country multiple cropping using HYV seeds, pesticides and large volume of surface and sub-surface water is practiced. It has already been discussed how it has created water pollution.

Water pollution affects the poor people most. Due to poverty it is not possible for them to enjoy the benefit of the modern water purifying technologies that the well-off people use. Very often poorer sections of the developing countries have to use rivers or other surface waterbodies, though polluted, due to absence of alternative sources. Being exposed to the pollutants they suffer from different diseases which make them expend a considerable portion of their little earnings. Failure of such treatment, mostly due to lack of money, brings death in apathy.

Conclusion

From the above discussion we can understand that water pollution is a serious problem for mankind. the pollutants not only have a short term effect but also have long term implications too. The profit maximization motive of production is mainly responsible for generation and increase of water pollution with time and over space. The entrepreneurs as well as the governments, who are basically run and supported by the entrepreneurs, are not interested to check pollution of rivers and other waterbodies, which are resources of the commons. In developed countries public consciousness has given rise to some green parties. The green parties by their green politics have, to some extent become successful in making governments work to check pollution. The technocrats have put forward various technological solutions in arresting pollutants. However, technological solutions have been formulated keeping in mind the market of the developed countries and not the poor millions of the developing countries. The developing countries are the worst sufferers from water pollution as we have seen in our discussion. Millions who suffer most from water pollution are cultivators, agricultural laborers, industrial workers and likewise who are the true producers of various resources and may be termed as the social autotrops. If they are being pushed close to the extinction then the entire human ecosystem is likely to collapse. Economic development of these people, spread of literacy and health consciousness among them should be given due importance along with efforts to provide them safe drinking water for drinking and other daily works. Fortunately various government and non-governmental organizations in different countries has put their effort to keep their environment clean and provide safe drinking water for their citizens. Sooner we get control of water pollution more secure will be our existence on this planet.

References

1. Bhardwaj, R.M. (2005). Water quality Monitoring in India–Achievements and Constraints. *IWG–Env, International Work Session on Water Statistics, Vienna, June 20–22.*
2. Bunker, S. (1984). Modes of extraction, unequal exchange, and the progressive underdevelopment of an extreme periphery: The Brazilian Amazon. *American Journal of Sociology*, 89: 1017–1064.
3. Hornborg, A. (2003). Cornucopia or zero-sum game? The epistemology of sustainability. *Journal of World–Systems Research* 9: 205–218.
4. Hornborg, A. (2001). *The Power of the Machine*. Walnut Creek: Alta Mira Press.
5. Jorgenson, A.K. (2006). Unequal ecological exchange and environmental degradation: A theoretical proposition and cross–national study of deforestation. *Rural Sociology*, 71: 685–712.
6. Jorgenson, A.K. and Rice, J. (2005). Structural dynamics of international trade and material consumption: A cross-national study of the ecological footprints of less-developed countries. *Journal of World–Systems Research*, 11: 57–77.
7. Karliner, J. (1997). *The Corporate Planet: Ecology and Economy in the Age of Globalization*. San Francisco: Sierra Club Books.
8. Manivasakam, N. (1984). *Environmental Pollution*. National Book Trust, Delhi, p. 41.
9. Rajagopalan, R. (2005). *Environmental Studies: From Crisis to Cure*. Oxford University Press, New Delhi.
10. Shandra, J.M., Shor, F. and London, B. (2009). World polity, unequal ecological exchange, and organic water pollution: A

cross-national analysis of developing nations. *Human Ecology Review*, Society for Human Ecology, 16(1): 53–63.

11. *The Economist*. 11 December 2008. "A special report on India: Creaking, groaning: Infrastructure is India's biggest handicap"
12. West, Larry (2006). March 26, "World Water Day: A Billion People Worldwide Lack Safe Drinking Water" *http: // environment.about.com*

Chapter 16

An Overview on Microbial Degradation of Petroleum Hydrocarbon Contaminants

Nitin Kumar and Anil Kumar

ABSTRACT

Microorganisms use to reduce/minimize the hazardous waste concentration from contaminated sites. Presently accepted disposal methods of incineration or burial insecure landfills have an expensive when amounts of contaminants are large. Mechanical and chemical methods generally used to remove hydrocarbons from contaminated sites are limited effective and expensive. But the Bioremediation is now used extensively for the biodegradation of hydrocarbon contamination resulting from the activities related to the petrochemical industry, which may refer to complete mineralization of organic contaminants into carbon dioxide, water, inorganic compounds, and cell protein or transformation of complex organic contaminants to other simpler organic compounds by biological agents like microorganisms. Many indigenous microorganisms in water and soil are capable of degrading hydrocarbon contaminants. This Chapter presents an updated overview of petroleum hydrocarbon degradation by microorganisms.

Keywords: *Microorganisms, Hydrocarbon, Bioremediation*

Introduction

Petroleum is a liquid mixture consisting of complex hydrocarbons. It is the main source of energy for industry and daily life. During the exploration, production, refining, transport, and storage of petroleum and petroleum products the amount of natural crude oil seepage was estimated to be 600,000 metric tons per year [1]. The main cause of water and soil pollution is the release of hydrocarbons into the environment whether accidentally or due to human activities [2]. Soil contamination with hydrocarbons causes extensive damage of local system since accumulation of pollutants in animals and plant tissue may cause death or mutations [3]. The technology commonly used for the soil remediation includes mechanical, burying, evaporation, dispersion and washing are expensive and can lead to incomplete decomposition of contaminants. The process of bioremediation defined as the use of biological agents, such as bacteria, fungi, or green plants, to remove or neutralize contaminants, as in polluted soil or water. Bacteria and fungi generally work by breaking down contaminants such as petroleum into less harmful substances [4]. It is a noninvasive and relatively cost-effective technology [5]. Oil spill bioremediation depends on presence of microorganisms with the appropriate metabolic capabilities. The optimal rates of growth and hydrocarbon biodegradation can be sustained by the adequate concentrations of nutrients, oxygen and pH values (6-9). Oil and oil surface area are also important factors for bioremediation success. Oil spill bioremediation can be done by two processes namely Bioaugmentation and Biostimulation. Bioremediation has been used on very large scale application, as demonstrated by the shore-line clean-up effortsin Prince William Sound, Alaska, after the Exxon Oil spill. Although the Alaska oil-spill cleanup represents the most extensive use of bioremediationon any one site, there have been many other successful application on smaller scale. This Chapter provides extensive information on microbial degradation of petroleum hydrocarbon contaminants towards the better understanding in bioremediation challenges.

Microbial Degradation of Petroleum Hydrocarbons

Rates of biodegradation depend greatly on the composition, state, and concentration of the oil or hydrocarbons, with dispersion and emulsification enhancing rates in aquatic systems and absorption by soil particulates being the key feature of terrestrial ecosystems. Hydrocarbon degradation may be influenced by different factors such as their limited availability to microorganisms. The susceptibility of hydrocarbons to microbial degradation can be generally ranked as follows: linear alkanes branched alkanes small aromatics cyclic alkanes [6,7]. Microbial degradation is one of the most promising mechanisms for removing the petroleum hydrocarbon pollutants from the environment [8-10]. There are different types of microorganisms such as bacteria, fungi and yeast, which are responsible for degradation of petroleum hydrocarbon pollutants. The microorganisms, namely, *Arthrobacter, Burkholderia, Mycobacterium, Pseudomonas, Sphingomonas, and Rhodococcus*were found to be involved for alkyl aromatic degradation in marine sediments which was reported by Jones *et al.* Individual population can't degrade complex mixture of hydrocarbon such as crude oil in soil, fresh water and marine environment, but mixed population with overall broad enzymatic capacities enhanced the rate. Bacteria are the most active and primary agents in petroleum degradation. Several bacteria are even known to feed exclusively on hydrocarbons [11]. Floodgate [12] listed 25 genera of hydrocarbon degrading bacteria and 25 genera of hydrocarbon degrading fungi, which were isolated from marine environment. Das and Mukherjee reported the crude petroleum oil from petroleum contaminated soil from North East India. There are various bacterial genera namely, *Gordonia, Brevibacterium, Aeromicrobium, Dietzia, Burkholderia* and *Mycobacterium* isolated from petroleum-contaminated soil, which have the capability for hydrocarbon degradation [15]. Some hydrocarbons such as poly-aromatic hydrocarbon may not be degraded at all but now it can be done by *Sphingomonas*. Fungal genera,namely, *Amorphoteca, Neosartorya, Talaromyces,* and *Graphium* and yeast genera, namely, *Candida, Yarrowia,* and *Pichia* were isolated from petroleum-contaminated soil and proved to be the potential organisms for hydrocarbon degradation [15].

Factors Influencing Petroleum Hydrocarbon Degradation

The fate of petroleum hydrocarbons in the environment is mainly determined by abiotic factors, which influence the Weathering, including biodegradation of the oil. Factors which influences rates of microbial growth and enzymatic activities affect the rates of petroleum pollutants depends on the quantity and quality of the hydrocarbon mixture and on the properties of the affected ecosystem. In one environment petroleum hydrocarbon can persist indefinitely, where as under another set of conditions the same hydrocarbons can be completely biodegraded within a relatively few hours or days. Brusseau discussed the number of limiting factors that affect the biodegradation of petroleum hydrocarbons [16]. Among physical factors, temperature plays an important role in biodegradation of hydrocarbons. Atlas [17] found that at low temperatures, the viscosity of the oil increased, while the volatility of the toxic low molecular weight hydrocarbons were reduced, delaying the onset of biodegradation. Temperature also affects the solubility of hydrocarbons [21]. Although hydrocarbon biodegradation can occur over a wide range of temperatures, the rate of biodegradation generally decreases with the decreasing temperature.

Nutrients (such as nitrogen, phosphorus and in rarely iron) also play a very main role in biodegradation of hydrocarbon pollutants. Some of these nutrients could become limiting factor thus affecting the biodegradation processes. Atlas [11] reported that when a major oil spill occurred in marine and freshwater environments, the supply of carbon was significantly increased and the availability of nitrogen and phosphorus generally became the limiting factor for oil degradation. In marine environments, it was found to be more pronounced due to low levels of nitrogen and phosphorous in seawater [12]. Freshwater wetlands are typically considered to be nutrient deficient due to heavy demands of nutrients by the plants [22]. Therefore, additions of nutrients were necessary to enhance the biodegradation of oil pollutant [23,24]. On the other hand, excessive nutrient concentrations can also inhibit the biodegradation activity [25]. Several authors have reported the negative effects of high NPK levels on the biodegradation of hydrocarbons [26,27] especially on aromatics [28].

Mechanism of Petroleum Hydrocarbon Degradation

The most rapid and complete degradation of the majority of organic pollutants is brought about under aerobic conditions. Oxygenases and Peroxidases are the main enzymes, which are used, in intracellular attack of inorganic pollutants and activation as well as incorporation of oxygen in the enzyme key reaction. Peripheral degradation pathways convert organic pollutants step by step into intermediates of the central intermediary metabolism, for example, the tricarboxylic acid cycle. Biosynthesis of cell biomass occurs from the central precursor metabolites, for example, acetyl-CoA, succinate, pyruvate. Sugars required for various biosyntheses and growths are synthesized by gluconeogenesis. The degradation of petroleum hydrocarbons can be mediated by specific enzyme system.

Cytochrome P450 alkane hydroxylases constitute a super family of ubiquitous Heme-thiolate Monooxygenases which play an important role in the microbial degradation of oil and other compounds. Higher eukaryotes generally contain several different P450 families that consist of large number of individual P450 form. Cytochrome P450 enzyme systems was found to be involved in biodegradation of petroleum hydrocarbons. Yeast species are capable to use n-alkanes and other aliphatic hydrocarbons as a source of C and energy and it is governed by multiple microsomal cytochrome P-450 forms isolated from yeast species such as *Candida maltosa, Candida tropicalis,* and *Candida apicola* .

Uptake of Hydrocarbon by Biosurfactants

Biosurfactants are heterogeneous group of surface-active chemical compounds produced by a wide variety of microorganisms [18-20, 31-33]. Surfactants enhance solubilization and removal of contaminants [34,35]. Biodegradation is also enhanced by surfactants due to increased bioavailability of pollutants. Bioremediation of oil sludge using biosurfactants has been reported by Cameotra and Singh [36]. *Pseudomonads* are the best known bacteria capable of utilizing hydrocarbons as carbon and energy sources and producing biosurfactants [13,36,37]. Among *Pseudomonads, P. aeruginosa* is widely studied for the production of glycolipid type biosurfactants. However, glycolipid type biosurfactants are also reported from some other species like *P. putida* and *P. chlororaphis*. Biosurfactants increase the oil surface area for bacteria to utilize it [38].

Commercially Available Bioremediation Agents

Microbiological cultures, enzyme additives, or nutrient additives that significantly increase the rate of biodegradation were defied as bioremediation agents by USEPA [39]. Bioremediation agents are classified as bioaugmentation agents and biostimulation agents based on the two main approaches to oil spill bioremediation. Numerous bioremediation products have been proposed and promoted by their vendors, especially during early 1990s, when bioremediation was popularized as "the ultimate solution" to oil spills [40]. Studies showed that bioremediation products may be effective in the laboratory but significantly less so in the field [41]. This is because laboratory studies cannot always simulate complicated real world conditions such as spatial heterogeneity, biological interactions, climatic effects, and nutrient mass transport limitations. Therefore, field studies and applications are the ultimate tests or the most convincing demonstration of the effectiveness of bioremediation products.

Phytoremediation and Genetically Modified Bacteria

Phytoremediation is an emerging technology that uses plants to manage a wide variety of environmental pollution problems, including the cleanup of soils and groundwater contaminated with hydrocarbons and other hazardous substances. The different mechanisms, namely, hydraulic control, phytovolatilization, rhizoremediation, and phytotransformation. Could be utilized for the remediation of a wide variety of contaminants. Phytoremediation can be cost-effective (a) for large sites with shallow residual levels of contamination by organic, nutrient, or metal pollutants, where contamination does not pose an imminent danger and only "polishing treatment" is required; (b) where vegetation is used as a final cap and closure of the site. Phytoremediation could be applied for the remediation of various contaminated sites. Genetically engineered microorganisms (GEMs) are useful to improve the degradation of hazardous wastes under laboratory conditions. The genetically engineered bacteria showed higher degradative capacity. However, ecological and environmental concerns and regulatory constraints are major obstacles for testing GEM in the field. These problems must be solved before GEM can provide an effective clean-up process at lower cost. The use of genetically engineered bacteria was applied to bioremediation process monitoring, strain monitoring, stress response, end-point analysis, and toxicity assessment.

Conclusion

- ✰ Bioremediation has high ecological significances that depend on the indigenous microorganisms to mineralize the organic contaminants.
- ✰ Microorganism has enzymes systems to degrade and utilize different hydrocarbons as a source of carbon and energy.
- ✰ Bioremediation is a more promising technology than mechanical, burying, evaporation dispersion and washing because these technologies are expensive and can lead to incomplete decomposition of contaminants.
- ✰ The use of genetically modified bacteria represents research frontier with broad implications.

References

1. Kvenvolden, K.A. and Cooper, C.K. (2003). Natural seepage of crude oil into the marine environment. *Geo–Marine Letters*, 23(3–4): 140–146.
2. Holliger, C., Gaspard, S., Glod, G., Heijman, C., Schumacher, W., Schwarzenbach, R.P. and Vazquez, F. (1997). Contaminated environments in the subsurface and bioremediation: Organic contaminants. *FEMS Microbiology Reviews*, 20(3–4): 517–523.
3. Alvarez, P.J.J. and Vogel, T.M. (1991). Substrate interactions of benzene, toluene, and para-xylene during microbial degradation by pure cultures and mixed culture aquifer slurries. *Applied and Environmental Microbiology*, 57(10): 2981–2985.
4. Medina-Bellver, J.I., Marín, P., Delgado, A., Rodríguez-Sánchez, A., Reyes, E., Ramos, J.L. and Marqués, S. (2005). Evidence for *in situ* crude oil biodegradation after the Prestige oil spill. *Environmental Microbiology*, 7(6): 773–779.
5. April, T.M., Foght, J.M. and Currah, R.S. (2000). Hydrocarbon-degrading filamentous fungi isolated from flare pit soils in northern and western Canada. *Canadian Journal of Microbiology*, 46(1): 38–49.
6. Ulrici, W. (2000). Contaminant soil areas, different countries and contaminant monitoring of contaminants. In: *Environmental*

Process II. Soil Decontamination Biotechnology, (Eds.) H.J. Rehm and G. Reed, 11: 5–42.

7. Perry, J.J. (1984). Microbial metabolism of cyclic alkanes. In: *Petroleum Microbiology*, (Ed.) R.M. Atlas. Macmillan, New York, NY, USA, pp. 61–98.
8. Atlas, R.M. (1992). Petroleum microbiology. InL *Encyclopedia of Microbiology*, Academic Press, Baltimore, Md, USA, pp. 363–369.
9. Amund, O.O. and Nwokoye, N. (1993). Hydrocarbon potentials of yeast isolates from a polluted Lagoon. *Journal of Scientific Research and Development*, 1: 65–68.
10. Lal, B. and Khanna, S. (1996). Degradation of crude oil by *Acinetobacter calcoaceticus* and*Alcaligenes odorans*. *Journal of Applied Bacteriology*, 81(4): 355–362.
11. Atlas, R.M. (1985). Effects of hydrocarbons on micro-organisms and biodegradation in Arctic ecosystems. In: *Petroleum Effects in the Arctic Environment*, (Ed.) F.R. Engelhardt. Elsevier, London, UK, pp. 63–99.
12. Floodgate, G. (1984). The fate of petroleum in marine ecosystems. In: *Petroleum Microbiology*, (Ed.) R.M. Atlas. MacMillan, New York, NY, USA, pp. 355–398.
13. Rahman, K.S.M., Rahman, T.J., Kourkoutas, Y., Petsas, I., Marchant, R. and Banat, I.M. (2003). Enhanced bioremediation of n-alkane in petroleum sludge using bacterial consortium amended with rhamnolipid and micronutrients. *Bioresource Technology*, 90(2): 159–168.
14. Yakimov, M.M., Timmis, K.N. and Golyshin, P.N. (2007). Obligate oil-degrading marine bacteria. *Current Opinion in Biotechnology*, 18(3): 257–266.
15. Chaillan, F., Le Flèche, A., Bury, E., Phantavong, Y.-H., Grimont, P., Saliot, A. and Oudot, J. (2004). Identification and biodegradation potential of tropical aerobic hydrocarbon-degrading microorganisms. *Research in Microbiology*, 155(7): 587–595.
16. Brusseau, M.L. (1998). The impact of physical, chemical and biological factors on biodegradation. In: *Proceedings of the*

International Conference on Biotechnology for Soil Remediation: Scientific Bases and Practical Applications, (Ed.) R. Serra. C.I.P.A. S.R.L., Milan, Italy, pp. 81–98.

17. Atlas, R.M. (1975). Effects of temperature and crude oil composition on petroleum biodegradation. *Journal of Applied Microbiology*, 30(3): 396–403.

18. Muthusamy, K., Gopalakrishnan, S., Ravi, T.K. and Sivachidambaram, P. (2008). Biosurfactants: properties, commercial production and application. *Current Science*, 94(6): 736–747.

19. Mahmound, A., Aziza, Y., Abdeltif, A. and Rachida, M. (2008). Biosurfactant production by *Bacillus* strain injected in the petroleum reservoirs," *Journal of Industrial Microbiology and Biotechnology*, 35: 1303–1306.

20. Ilori, M.O., Amobi, C.J. and Odocha, A.C. "Factors affecting biosurfactant production by oil degrading *Aeromonas* spp. isolated from a tropical environment.

21. Foght, J.M., Westlake, D.W.S., Johnson, W.M. and Ridgway, H.F. (1996). Environmental gasoline-utilizing isolates and clinical isolates of *Pseudomonas aeruginosa* are taxonomically indistinguishable by chemotaxonomic and molecular techniques. *Microbiology*, 142(9): 2333–2340.

22. Mitsch, W.J. and Gosselink, J.G. (1993). *Wetlands*, 2nd edn. John Wiley and Sons, New York, NY, USA.

23. Choi, S.-C., Kwon, K.K., Sohn, J.H. and Kim, S.-J. (2002). Evaluation of fertilizer additions to stimulate oil biodegradation in sand seashore mesocosms. *Journal of Microbiology and Biotechnology*, 12(3): 431–436.

24. Kim, S.-J. Choi, D.H., Sim, D.S. and Oh, Y.-S. (2005). Evaluation of bioremediation effectiveness on crude oil-contaminated sand. *Chemosphere*, 59(6): 845–852.

25. Chaillan, F., Chaîneau, C.H., Point, V., Saliot, A. and Oudot, J. (2006). Factors inhibiting bioremediation of soil contaminated with weathered oils and drill cuttings. *Environmental Pollution*, 144(1): 255–265.

26. Oudot, J., Merlin, F.X. and Pinvidic, P. (1998). Weathering rates of oil components in a bioremediation experiment in estuarine sediments. *Marine Environmental Research*, 45(2): 113–125.

27. Chaîneau, C.H., Rougeux, G., Yéprémian, C. and Oudot, J. (2005). Effects of nutrient concentration on the biodegradation of crude oil and associated microbial populations in the soil. *Soil Biology and Biochemistry*, 37(8): 1490–1497.

28. Carmichael, L.M. and Pfaender, F.K. (1997). The effect of inorganic and organic supplements on the microbial degradation of phenanthrene and pyrene in soils. *Biodegradation*, 8(1): 1–13.

29. Chaillan, F., Le Flèche, A., Bury, E., Phantavong, Y.H., Grimont, P., Saliot, A. and Oudot, J. (2004). Identification and biodegradation potential of tropical aerobic hydrocarbon-degrading microorganisms. *Research in Microbiology*, 155(7): 587–595.

30. Scheuer, U., Zimmer, T., Becher, D., Schauer, F. and Schunck, W.-H. (1998). Oxygenation cascade in conversion of n-alkanes to α,ω-dioic acids catalyzed by cytochrome P450 52A3. *Journal of Biological Chemistry*, 273(49): 32528–32534.

31. Ilori, M.O., Adebusoye, S.A. and Ojo, A.C. (2008). Isolation and characterization of hydrocarbon-degrading and biosurfactant-producing yeast strains obtained from a polluted lagoon water. *World Journal of Microbiology and Biotechnology*, 24(11): 2539–2545.

32. Kiran, G.S., Hema, T.A., Gandhimathi, R., Selvin, J., Thomas, T.A., Rajeetha Ravji, T. and Natarajaseenivasan, K. (2009). Optimization and production of a biosurfactant from the sponge–associated marine fungus *Aspergillus ustus* MSF3. *Colloids and Surfaces B*, 73(2): 250–256.

33. Obayori, O.S., Ilori, M.O., Adebusoye, S.A., Oyetibo, G.O., Omotayo, A.E. and Amund, O.O. (2009). Degradation of hydrocarbons and biosurfactant production by *Pseudomonas* sp. strain LP1. *World Journal of Microbiology and Biotechnology*, 25(9): 1615–1623.

34. Brusseau, M.L., Miller, R.M., Zhang, Y., Wang, X. and Bai, G.Y. (1995). Biosurfactant and cosolvent enhanced remediation of contaminated media. *ACS Symposium Series*, 594: 82–94.

35. Bai, G., Brusseau, M.L. and Miller, R.M. (1997). Biosurfactant–enhanced removal of residual hydrocarbon from soil. *Journal of Contaminant Hydrology*, 25(1–2): 157–170.

36. Cameotra, S.S. and Singh, P. (2008). Bioremediation of oil sludge using crude biosurfactants. *International Biodeterioration and Biodegradation*, 62(3): 274–280.

37. Pornsunthorntawee, O., Wongpanit, P., Chavadej, S., Abe, M. and Rujiravanit, R. (2008). Structural and physicochemical characterization of crude biosurfactant produced by *Pseudomonas aeruginos* SP4 isolated from petroleum–contaminated soil. *Bioresource Technology*, 99(6): 1589–1595.

38. Nikolopoulou, M. and Kalogerakis, N. (2009). Biostimulation strategies for fresh and chronically polluted marine environments with petroleum hydrocarbons. *Journal of Chemical Technology and Biotechnology*, 84(6): 802–807.

39. Nichols, W.J. (2001). The U.S. Environmental Protect Agency: National Oil and Hazardous Substances Pollution Contingency Plan, Subpart J Product Schedule (40 CFR 300.900). In: *Proceedings of the International Oil Spill Conference*, pp. 1479–1483, American Petroleum Institute, Washington, DC, USA.

40. Hoff, R.Z. (1993). Bioremediation: an overview of its development and use for oil spill cleanup. *Marine Pollution Bulletin*, 26(9): 476–481.

41. Lee, K., Tremblay, G.H., Gauthier, J., Cobanli, S.E. and Griffin, M. (1997). Bioaugmentation and biostimulation: A paradox between laboratory and field results. In: *Proceedings of the International Oil Spill Conference*, pp. 697–705, American Petroleum Institute, Washington, DC, USA.

Chapter 17

Drinking Water Quality in Tribal Areas of Santal Parganas: Issues and Approaches

P.K. Verma, N.K. Mandal, Rishikesh Kumar, C.S. Azad and Chandan Kumar

- ☆ Safe water is essential for human well being.
- ☆ Supply of safe drinking water is a constitutional mandate, with Article 47.
- ☆ Integration of human health and water quality is a challenge and of growing concern.
- ☆ Surface water sources are not acceptable for drinking purpose as these are often loaded by various organic, inorganic and biological pollutant.
- ☆ Undergroundwater is the only alternate as safe, secured and palatable water.

Factors Causing Water Quality Problems

- ☆ Improper sewage discharge
- ☆ Open defecation
- ☆ Disposal of solid wastes
- ☆ Runoff from agricultural fields
- ☆ Over exploitation
- ☆ Chemical pollution
- ☆ Microbial pollution

All these factors may create health hazards

Objective

- ☆ To assess the quality problems and its impact on people health.
- ☆ To strengthen community based water quality monitoring programme.
- ☆ To review water quality standards based on scientific research.
- ☆ To recommend certain effective measures to ensure safe, affordable and sustainable potable water for common people.

Investigation Area

Tribal blocks-Sunderpahari and Boarijore of Godda district under Santal Pargana of Jharkhand.

The Primitive and pioneer community of this area are Paharias and Santhals. They are any how living amidst various natural and man-made problems with acute scarcity of drinking water, limited water resources, poor general health and lack of sanitation.

The investigation conducted to analyse physico-chemical and biological properties of different potable water sources available in tribal areas.

This include open dug well, tube well and hill stream flowing across tribal villages in bushy vegetation of hilly forest areas.

Methodology

- ☆ On spot observation of investigation area.
- ☆ Survey with appropriate village schedule, map and sampling kit.
- ☆ Laboratory examination for water analysis.

Groundwater with Chemical and Microbial Pollution: The Emerging Challenge

- ☆ Groundwater is used extensively and accounts for 90 per cent of the domestic water need in rural areas.
- ☆ Generally less susceptible to contamination and pollution.

- Natural impurities in rain water are removed during infiltration through soil strata.
- Variety of land and water based human activities are causing pollution of this precious resource.
- Open dug well are subjected to contamination due to unhygienic situation around the well and human interference in handling and extraction of water.
- Tube well water is often contaminated due to soil texture and infiltration rate that affects the downward movement of pollutant due to seepage.
- Geo-hydro chemical processes activated by pumping help in leaching of solid liquid wastes and toxic chemicals from earth rock and subsoil.
- ISI, ICMR, WHO recommended the standards for drinking water accepted by the Ministry of Health, Govt. of India.
- High turbidity and Acidic medium are responsible for wide spread gastric disorder and water borne diseases. Water is of soft category.
- PO_4^{3-} content exceeded the prescribed limit probably due to rainwater percolation, leaching of subsoil, surface soil, weathering of rocks and human activities.
- Higher values of nitrate in hill stream, due to influx of organic matter, improper sanitation and waste disposal etc.
- All water sources depicted bacterial contamination is a major concern, probably due to lack of sanitary protection around well and tube well, seepage of stagnant water.
- Heavy metal contamination for chromium and selenium recoded above the permissible limit in dugwell and hill stream.

Biomagnification of Such Metals in Food Chain and its Drastic Effect on the Organisms can not be Overlooked

Major Findings

- Bacterial contamination especially faecal coliform in water is a widesdpread problem in this area.

Table 17.1: Water Quality Problems and Associated Health Implications

Parameter	*Maximum Permissible Limit*	*Health Impact*
Fluoride	1.5mg/l	Dental fluorosis, skeletal fluorosis with severe bone damage, mental retardation, digestive disorder.
Arsenic	0.05mg/l	Warts and nodules on hands and feet, keratosis and skin lesions, change in skin pigmentation-long term exposure. Vomiting, oesophageal and abdominal pain diarrhoea are immediate symptoms.
Iron	1 mg/l	Digestive disorder, skin diseases, damage of blood cells, poisoning effect on child.
Nitrate	20 mg/l	Methaemoglobinemia (Blue body disease) due to decrease efficiency of hemoglobin to combine with oxygen
Selenium	0.01mg/l	Inflammation in gastro-intestinal tract, long term effects on kidney, liver and lungs.
Chromium	0.05mg/l	Carcinogenic effect in long term exposure.
Pesticides (Aldrin, Endosulfan etc.)		Weakend immunity, abnormal multiplication of cells leading to tumour formation. It may cause reproductive and endocrine damage, suspected carcinogen.

Table 17.2: Evaluation of Physico-Chemical and Biological Parameters with Standard Sets for Potable Water (Range Value)

Parameters	*Water Resources*			*Prescribed Standard ISI, ICMR, WHO*
	Dug well	*Tube Well*	*Hill Stream*	
Turbidity (NTU)	15-50	10-16	10-26	5-25
Conductivity (µs/cm)	200-230	200-275	125-90	1600
TDS (ppm)	140-260	160-300	270-325	500-1500
pH	5.6-6.6	6.2-6.8	5.2-6.7	6.5-8.5
DO ppm)	4.5-6.7	4.1-5.6	6.4-10.0	4.0-6.0
FCO_2 (ppm)	3.0-5.0	8.0-9.5	3.5-8.0	6
Total Hardness (ppm)	15-34	14-31	33-68	500
HCO_3	84-120	105-165	115-170	600
Chloride (ppm)	40-74	42-56	41-55	250
NO_3 (ppm)	0.46-0.70	0.45-0.66	0.66-0.89	20
PO_4 (ppm)	0.50-0.72	0.40-0.68	0.15-0.72	0.1
Total Coliform (MPN/100ml)	146-300	12-120	800-3800	0-50
Faecal coliform (MPN/100ml)	09-110	14-Jul	410-940	–
Cu (ppm)	0.003-0.005	0.015-0.017	0.022-0.28	0.05-1.0
Zn (ppm)	0.200-0.300	0.079-0.200	0.065-0.090	5.0-15.0
Ni (ppm)	0.014-0.016	0.002-0.003	0.002-0.004	<0.05
Co (ppm)	0.002-0.003	0.002-0.006	0.002-0.006	–
Pb (ppm)	0.005-0.009	0.003-0.005	0.003-0.006	0.05-0.10
Mn (ppm)	0.013-0.019	0.004-0.009	0.007-0.015	01-1.0
Fe (ppm)	0.007-0.009	0.014-0.016	0.008-0.015	0.1-1.0
Cd (ppm)	0.014-0.019	0.005-0.008	0.013-0.019	0.005-0.01
Cr (ppm)	0.038-0.066	BDL	0.031-0.069	0.05
Se (ppm)	0.040-0.090	0.014-0.016	0.040-0.080	0.01-0.05
As (ppm)	BDL	BLD	BDL	

- ☆ The area is identified as prone to several water borne diseases like diarrhoea, typhoid, Jaundice, gastro-enteritis in terms of health standard of the poor people.
- ☆ Open defecation, poor system of solid waste and wastewater disposal, poor maintenance of handpump and platform and unattended animal wastes constitute fairly high proportion of sanitary risk of groud water sources and hill stream.
- ☆ Heavy metal contamination of Cr and Se which stands slightly above the prescribed standard may have long-term effect through biomagnifications.
- ☆ Lack of technical knowledge about the handling and storage of water.

Remedial Measures

- ☆ Need for Community involvement in water quality management to ensure safe, good quality and adequate drinking water supply. Regular water quality monitoring, data processing, evaluation, sanitary monitoring and hygiene education.
- ☆ Protection and care of perennial/seasonal hill streams.
- ☆ Existing dug wells need renovation.
- ☆ Installation of hand pump, tube well at suitable place in functional order.
- ☆ Simple and low cost water treatment technology at household level is required.

  Morenga olifera (drum stick) seeds inhibit growth of bacteria and fungi in water.

  Osmium sanctum (tulsi) leaf act as water purifier with antibacterial and insecticidal properties.
- ☆ Community awareness drive and role of Panchayats to sustain decentralized water quality monitoring process to provide low cost training on water quality and setting up water quality lab at village/block level would facilitate the information flow and mass awareness.
- ☆ To provide low cost water testing kit to create interest and awareness among village students and people at large.

- ☆ Need to harvest, conserve and utilize rain water in appropriate manner by *Recharge pits, Recharge trenches, Dugwell, Injection well* and *percolation tank* to solve the crisis.
- ☆ *Accountability: Users* should realize the individual responsibility in maintaining the quality of water. Factors like contamination at source and storage in clean and covered vessels lies with the users.
- ☆ Wastewater treatment and its reuse to reduce the burden on fresh water sources.

References

1. A.P.H.A. (1975). *Standard Methods for the Examination of Water and Wastewater*, 14th edn. American Water Works Association, Washington, New York, pp. 1193.
2. APHA, AWWA, WPCA (1998). *Standard Methods for the Examination of Water and Wastewater*, 20th edn. American Public Hlth. Assoc., Washington, USA.
3. Ganapati, S.V. (1943). An ecological survey of a garden pond containing abundant zooplankton. *Proc. Indian Acad. Sci. (B)*, 17: 41–58.
4. Goel, P.K., Trivedi, R.K. and Bhave, S.V. (1985). Studies on the limnology of a few fresh water bodies in South Western Maharastra. *Indian J. Envi. Pro.*, 5(1): 19–25.
5. Golterman, H.L., Clymo, R.S. and Ohnstad, M.A.N. (1987). *Methods for Physical and Chemical Analysis of Freshwaters*, 2nd edn. I.B.P. Manual. Blackwell Scientific Publ., Oxford, UK.
6. Hussainy, S.U. (1965). Limnological studies of the departmental pond of Annamalainagar. *Environ. Health*, 7: 24–37.
7. Kumar, A. and Siddiqui, E.N. (1997). Quality of drinking water in and around Ranchi. *Indian J. Environ. Prol.*, 18(5): 339–345.
8. Kumar, A. and Verma, P.K. (2001). Ecological status of Masanjore Reservoir in relation to fisheries management in Santhal Pargana (Jharkhand), India. In: *Ecol. and Conservation of Lakes, Reservoirs and Rivers*, (Ed.) A. Kumar. ABD Publishers, Jaipur, pp. 889.
9. Kumaran, P. (1966). A natural sewage stabilization pond at Jaipur. *Environmental Health*, 8: 134–141.

10. Michael, R.G. (1964). Diurnal variation of the plankton correlated with physico-chemical factors in three different ponds. *Ph.D. Thesis,* CalcuttaUniversity, Calcutta.

11. Michael, R.G. (1969). Seasonal trends in physico-chemical factors and plankton of freshwater fish pond and their role in fish culture. *Hydrobiologia,* 33(1): 144–160.

12. NEERI (1986). *Manual on Water and Wastewater Analysis.* National Environmental Engineering Research Institute, Nagpur, pp. 340.

13. Saha, L.C. (1985). Changes in the properties of bottom soil of two freshwater ponds in relation to ecological factors. *Indian J. Ecol.,* 12(1): 147–150.

14. Saha, L.C. and Pandit, B. (1984). Comparative ecology of Bhagalpur ponds and river Ganges during summers. *Nat. Acad. Sci. Letters,* 7(10): 295–296.

15. Trivedi, R.K. and Goel, P.K. (1986). *Chemical and Biological Method for Water Pollution Studies.* Environment publications Karad, India.

16. Verma, M.N. (1967). Diurnal variation in a fish pond in Seoni, India. *Hydrobiologia,* 30(1): 129–137.

Chapter 18

Controlling the Environmental Pollution Eco-friendly through Microbes and Biotechnology

Jahangeer and Anil Kumar

ABSTRACT

The most industrially important primary metabolites are the amino acids, nucleotides, vitamins, solvents, and organic acids and their microbial biosynthesis is versatile. Millions tons of amino acids are produced every year with a total multibillion dollar market. Microbes contribute to geo-chemical cycles in the ecosystem, which contributes in biodegradation and bioremediation of contaminated environments, and have a great potential in energy conversion and regeneration. Presently, at least 150 genomes of non-pathogenic microbes have been sequenced, mostly are bacteria from various environments. Soil is a repository of diverse microorganisms, which has frequently been used to isolate and exploit microbes for industrial, environmental and agricultural applications. Microbes can biodegrade organic chemicals and purposeful enhancement of this natural process can aid in pollutant degradation and waste-site cleanup operations. The emerging field 'metagenomics' in combination with the high-throughput sequencing technology offers opportunities to discover new functions of microbes in the environment on a large scale, and has become the 'hot spot' in the field of environmental microbiology. The functional genomics is currently the most effective approach for increasing the knowledge at the molecular level of metabolic and adaptive processes in whole cells. High-throughput technologies, such as DNA microarrays, and improved two-dimensional electrophoresis methods combined with tandem mass-spectroscopy, supported by bio-informatics, are useful tools for development of genetically engineered microbes, which are used for pollutant degradation. This Chapter deals the functions of microbes and biotechnology in the development of suitable environmental condition for living organisms.

Keywords: *Microbes, Biodegradable pollutant, Biotechnology and Genomics.*

Introduction

The paramount of pollution in our environment is a dire consequence of continually expanding population along with an exponential development in the industrial field. Microbes are ubiquitous in nature and are being exposed to the continuous release of more and more recalcitrant xenobiotic compounds into the environment [1]. No wonder, these microbes, inhabiting polluted environments, are armed with various resistance and catabolic potentials. The catalytic potential of microbes in nature is enormous and this is advantageous to mankind for a cleaner and healthier environment through bioremediation [2]. In general, potential microbes with broad spectrum of activities from their native habitat have been screened, characterized, genetically modified and released back to their native habitat for better performance. By such studies, the core problem of pollution is tactfully tackled and benefits of decontamination add healthy atmosphere to mankind. The purified degrading enzymes, Nitrilase, Azoreductases and Oragnophosphate hydrolases could be effectively used in industry for the treatment of effluents [3]. The systems developed are eco-friendly and economical and hence could effectively be integrated with physico-chemical methods for pollution control. The index of xenobiotic compounds released into the environment increases due to industrialization and combating pollution by the release of these compounds is essential for the sustenance of the future generation. In this context, microbes such as algae, fungi and bacteria, play an important role by giving us a helping hand in bioremediation of these xenobiotic compounds. Degradation of pesticides by different bacterial population proves to be the best example for citing the role of microbes in bioremediation of xenobiotic compounds [4]. A large number of pesticides and insecticides like morpholine, methyl parathion, organophosphorous compounds and benzimidazoles are widely used to increase the agricultural output and has also contributed to the pollution load, as many of these man-made chemicals are non-biodegradable [5]. The pollution control strategies involving physico-chemical methods many a time aggravate the problem, rather than eliminating it. Microbes play a very important role in the mineralization of pollutants either by natural selection or through recombinant DNA technology making bioremediation process an extension of normal microbial metabolism [6,7]. Xenobiotic compounds are also widely employed in our day-to-day life. Microbes also mediate degradation of xenobiotic compounds like dyes and plastics [8]. With the help of Biotechnology understanding the molecular biology of the microorganisms, and the ability to genetically manipulate the microorganisms and infuse engineering principles into biology have led to novel strategies for combating environmental problems. Construction of strains with broad spectrum of catabolic potential with heavy metal resistant traits makes them ideal for bioremediation of polluted environments in both aquatic and terrestrial ecosystems. The transfer of genetic traits from one organism to another paves way in creating Genetically Engineered Organisms (GEM's) for combating pollution in extreme environments making it a boon to mankind to cleanup the mess that has created in nature.

Microbes and Environmental Biotechnology

Environmental biotechnology is the use of living organisms for a wide variety of applications in hazardous waste treatment and pollution control. For example, a fungus is being used to clean up a noxious substance discharged by the papermaking industry [9]. Other naturally occurring microbes that live on toxic waste dumps are degrading wastes, such as polychlorinated biphenyls (PCBs), to harmless compounds. Marine biotechnologists are studying ways that estuarine bacteria can detoxify materials such as chemical sea brines that cause environmental problems in many industries [10]. Environmental biotechnology can more efficiently clean up many hazardous wastes than conventional methods and greatly reduce our dependence for waste cleanup on methods such as incineration or hazardous waste dumpsites.

How Does it Work?

Using biotechnology to treat pollution problems is not a new idea. Communities have depended on complex populations of naturally occurring microbes for sewage treatment for over a century. Every living organism-animals, plants, bacteria and so forth-ingests nutrients to live and produces a waste byproduct as a result. Different organisms need different types of nutrients [11]. Certain bacteria thrive on the chemical components of waste products. Some microorganisms, for example, feed on toxic materials such as methylene chloride, detergents and creosote. Bioremediation can be applied to recover brown fields for development and for preparing contaminated industrial effluents prior to discharge into waterways [12]. Ecologically microbes are most important for the restoration of contaminated environment. The processes of bioremediation is an economical, versatile, environment friendly and efficient treatment strategy, and a rapidly developing technology to degrade and or detoxify chemical substances such as petroleum products, aliphatic and aromatic hydrocarbons, industrial discharges, pesticides and their metabolites and metals [11,12,13,] Because of the biodiversity of microbes in nature, they can able to degrade variety of chemical and toxic pollutants. Hence the effects of contamination on environment can be minimized to larger extend. Bioremediation technologies are also applied to contaminated wastewater, ground or surface waters, soils, sediments and air where there has been either accidental or intentional release of pollutants or chemicals that pose a risk to human, animal or ecosystem health. Different approaches to bioremediation take advantage of the metabolic processes of different organisms for degradation, or sequestering and concentration, of different contaminants. Bioremediation using genetically engineered microorganisms (GEMs, or GMOs), carrying recombinant proteins, is still relatively uncommon due to regulatory constraints related to their release and control [15]. Other methods of enzyme optimization that do not include gene cloning techniques, might be applied to indigenous microorganisms in order to enhance their pre-existing traits. Environmental engineers use bioremediation in two basic ways. They introduce nutrients to stimulate the activity of bacteria already present in the soil at a hazardous waste site, or they add new bacteria to the soil. The bacteria then "eat" the hazardous waste at the site and turn it into harmless byproducts. After the bacteria consume the waste materials, they die off or return to their normal population levels in the environment. The vast majority of bioremediation applications use naturally occurring microorganisms to identify and filter manufacturing waste before it is introduced into the environment or to clean up existing pollution problems. In some cases, the byproducts of the pollution-fighting microorganisms are themselves useful. Methane, for example, can be derived from a form of bacteria that degrades sulfur liquor, a waste product of paper manufacturing. Biotechnology will also have an impact on two sources of energy: fossil fuels and new biomass-based fuels. Innovations wrought by biotechnology can help remove the sulfur from fossil fuels, significantly decreasing their polluting power [13,14]. Using biomass for energy has the same environmental advantages as using biomass feedstock's, so government labs have devoted significant resources to research on recombinant technology and bioprocess engineering to improve the economic feasibility of biomass-derived energy. Biotechnology developed biofuels, biocatalyst and green plastic these are eco-friendly product and have very less pollution as on date the most reliable strategy is biodegradation by eco-friendly microbes, which is generally accepted as an environmentally sound and economically feasible protocol for the treatment of hazardous waste and effluents.

The techniques of biotechnology are providing us with novel methods for diagnosing environmental problems and assessing normal environmental conditions so that we can be better-informed environmental stewards.

Beneficial Industries

- ☆ *The chemical industry*: Using biocatalysts to produce novel compounds, reduce waste byproducts and improve chemical purity.
- ☆ *The plastics industry*: Decreasing the use of petroleum for plastic production by making "green plastics" from renewable crops such as corn or soybeans.
- ☆ *The paper industry*: Improving manufacturing processes, including the use of enzymes to lower toxic byproducts from pulp processes.
- ☆ *The textiles industry*: Lessening toxic byproducts of fabric dying and finishing processes. Fabric detergents are becoming more effective with the addition of enzymes to their active ingredients.

- *The food industry*: Improving baking processes, fermentation-derived preservatives and analysis techniques for food safety.
- *The livestock industry*: Adding enzymes to increase nutrient uptake and decrease phosphate byproducts.

Biotechnology and its applications are diverse with the importance reflected in our daily lives and encompass other branches of science like medicine and agriculture.

Biotechnology and its Applications in the Benevolence of Mankind

- Biotechnology and its applications play important role in agriculture. It provides high yielding varieties of crops with lesser use of chemical pesticides utilizing bio-pesticides.
- Biotechnology aims to improve animal breeding techniques and breed selection of livestock. This also includes the development of vaccines and disease detection strategies contributing to the improvement of livestock in general.
- Biotechnology and its applications play role in industries as in development of biocatalysts. In this process, whereby the living organisms can be modified to produce biocatalysts called enzymes enabling their synthesis in commercial amounts.
- Biotechnology and its applications help better waste management treatment and pollution control strategies. Bioremediation is the significant biotechnology application whereby the microbes and their enzymes are utilized in order to recover the pollution- altered environment and also in recycling processes.
- Biotechnology helps to develop diverse pharmaceutical products for the treatment of wide range of human diseases. The examples include the development of human insulin to treat diabetes, human growth hormone, etc. Human genome sequencing project has paved the doorway to new therapeutic strategies.
- Biotechnology and its applications also aim for the

sustainable growth of aquaculture, apiculture, lac culture including the use of synthetic hormones in induced breeding, transgenic fish, gene banking etc.

- ☆ Biotechnology targets at production of various genetically modified organisms (GMOs) used for different benevolent usage.

Biotechnology and its Applications as Threat to the Mankind

- ☆ The horizontal gene transfer strategy used in biotechnology may produce new vectors that do not naturally exist.
- ☆ Biotechnology and its applications tend to increase the homozygosity in a population resulting in transformations of wild species and loss of biodiversity.
- ☆ Biotechnology and its applications can upset the natural ecological balance through release of genetically modified microbes.
- ☆ The dangers include the ethical and moral issues of cloning and bio-weapons.

Conclusions

- ☆ Biotechnology has a high ecological significance that depends on the indigenous microorganism to mineralize the organic contaminants.
- ☆ Microorganisms have enzymes systems to degrade and utilize different hydrocarbons as a source of carbon and energy.
- ☆ Biotechnology is a more promising technology than mechanical, burying, evaporation dispersion and washing because these technologies are expensive and can lead to incomplete decomposition of contaminants.
- ☆ The use of genetically modified bacteria represents research frontier with broad implications.

References

1. Alexander, M. (1999) *Biodegradation and Bioremediation*. Elsevier Science.

2. Diaz, E. (Ed.) (2008). *Microbial Biodegradation: Genomics and Molecular Biology*. Caister Academic Press.

3. Water and Environmental Health at London and Loughborough (1999). "*Wastewater Treatment Options*."Technical brief no. 64. London School of Hygiene and Tropical Medicine and Loughborough University.

4. Felsot, A., Maddox, J.V. and Bruce, W. (1981). Enhanced microbial degradation of carbofuran in soils with histories of Furadan use. *Bull. Environ. Contam. Toxicol.*, 26(6): 781–788.

5. Wentzel, M.C., Mbewe, A. and Ekama, G.A. (1995). Batch test for measurement of readily biodegradable COD and active organism concentrations in municipal wastewaters. *Water SA*, 21: 117–124.

6. Gallardo, M.E., Ferrandez, A., de Lorenzo, V., Garcia, J.L. and Diaz, E. (1997). Designing recombinant Pseudomonas strains to enhance biodesulfurization. *J. Bacteriol.*, 179: 7156–7160.

7. Chen, S. and Wilson, D.B. (1997). Construction and characterization of *Escherichia coli* genetically engineered for bioremediation of Hg^{2+}–contaminated sediments. *Appl. Environ. Microbiol.*, 63: 2442–2445.

8. Meyer, A. and Panke, S. (2008). Genomics in metabolic engineering and biocatalytic applications of the pollutant degradation machinery. *Microbial Biodegradation: Genomics and Molecular Biology*. Caister Academic Press.

9. Holliger, C., Gaspard, S., Glod, G., Heijman, C., Schumacher, W., Schwarzenbach, R.P. and Vazquez, F. (1997). Contaminated environments in the subsurface and bioremediation: organic contaminants. *FEMS Microbiology Reviews*, 20(3–4): 517–523.

10. Medina-Bellver, J.I., Marín, P., Delgado, A., Rodríguez-Sánchez, A., Reyes, E., Ramos, J.L. and Marqués, S. (2005). Evidence for *in situ* crude oil biodegradation after the Prestige oil spill. *Environmental Microbiology*, 7(6): 773–779.

11. April, T.M., Foght, J.M. and Currah, R.S. (2000). Hydrocarbon-degrading filamentous fungi isolated from flare pit soils in northern and western Canada. *Canadian Journal of Microbiology*, 46(1): 38–49.

12. Ulrici, W. (2000). Contaminant soil areas, different countries and contaminant monitoring of contaminants. In: *Environmental Process II: Soil Decontamination Biotechnology*, (Eds.) H.J. Rehm and G. Reed. 11: 5–42.

13. Amund, O.O. and Nwokoye, N. (1993). Hydrocarbon potentials of yeast isolates from a polluted Lagoon. *Journal of Scientific Research and Development*, 1: 65–68.

14. Lal, B. and Khanna, S. (1996). Degradation of crude oil by *Acinetobacter calcoaceticus* and *Alcaligenes odorans. Journal of Applied Bacteriology*, 81(4): 355–362.

15. Watanabe, K. and Kasai, Y. (2008). Emerging technologies to analyze natural attenuation and bioremediation. *Microbial Biodegradation: Genomics and Molecular Biology*. Caister Academic Press.

Chapter 19

Relationship between Economic Growth and Environment

Nitesh Raj and Hashmat Ali

ABSTRACT

The human beings of 21st century are surrounded in fire and their main problem is of selection between growth and environment. Both have a question of their existence. And the answer is hidden in co-operation between the food substance for increasing population and drinking water facilities, plenty of goods in consumption and lack of pure air (oxygen), expansion of irrigated land with contraction of land due to salt consolidation in fertile land and problem of water logging in irrigated land, degradation of food supply source in order to fulfill the supply of fuel and danger flooding of ice land and costal area due to global warming.

Keywords: *Environment, Productivity, Globalization, Pollution, Uranium, Nexus, Consumption, Growth; Economy, Materialistic life.*

Introduction

Man is a bigger enemy of himself. In order to see the physical growth he has distorted the environment. Now, in between the increasing economic growth and environment, pollution has become an important topic for discussion at local, national and international levels. The physical life standard of developed nations after 2nd world war has been increased tremendously but it also caused acid rain, lack of pure water, soil and air. The capital diseases like Cancer, AIDS, Hepatitis B and C also increased in the same proportion. In short, the physical resources are in plenty but there is scarcity of healthy and natural environment. Today the developed nations like USA, Canada, European Union, Australia, Japan which having per capital income more than $20000 per annum are facing severe environmental problems. People therefore are looking for healthy environmental places like costal areas, mountains, forests etc. But in all these places the environment is polluted due to heavy clot of vehicles, aeroplanes and the pollutants spread by people who are going to these places. The water is polluted due to petroleum spread in a considerable amount in Sea and in Beaches. The radioactive element is being spread in peripheral of Japan due to earthquake in 2011 despite the high security zone of atomic reactors and same condition is with Chalurvik of Chernobyl in Ukraine. India is facing a big problem in Jadugoda mines of Uranium where the radio active element has disturbed the socio-economic and environmental conditions of people. According to National Science Academic of USA "The blended form of industrial development can bring dangerous consequences for environment. It arises in the wake of utilising the traditional technologies by the underdeveloped countries. But the modern technologis are also not very effective to stop the environmental pollution. The acid rain disturbing water, soil and forest and the continuous expansion of the hole in ozone layer are its examples." According to World Resource Institute, there are two fundamental resources for environmental pollution or ecological degradation 1) Increase in pollution and 2) increase in consumption level. Besides this there are many political and economical resources among which important being market forces, government subsidy, globalization of product and market and government corruption. These are the factors which show us that people are utilizing the recourses recklessly. This is also determined by poverty level, land ownership and conditions of war and peace. In the next 50 years the population can increase up to 10 billion. How much it will tell upon the environment is just unimaginable. Presently the consumption level has increased many times compared to increase in population. During the period 2010-11, the world economy has changed by 3 times (population growth 25 per cent). The effect of globalization has increased the consumption level but this is creating a bad effect upon the ecological consumption cost.

For example the workman related to water does not know the resource of copper. During the process of excavation of copper, daily 80,000 tone wastage are thrown to Octedy River in Papua Newguini. This results in the death of various aguatic lives. The Wopkaimin fish which was the basis of livelihood for fishermen are also destroyed. The countries that are mainly responsible for environmental pollution are the developed country. There is also a difference in consumption level between developed country and under developed country. In USA in order to fulfill the consumption level of one person there is a need of 5 hectare ecological production system, while the availability of land in an underdeveloped country is only 0.5 hectare. In industrial country per head CO_2 spread is 11,000kg per year while in Asia it is 3000 kg per year. In a developed country a person spends approx $16000 per year in international market for his personal consumption, while in Asia it is $350 only.

Form above statements it is obvious that economic growth and environment are interdependant. Before 1992 in the Rio earth summit it was resolved that for development and poverty alleviation the environmental degradation is necessary but in 1997, in the Bruteland conference which was based upon Our Common Future it was stressed that ecological integrity in a sustainable condition is unavoidable and infact healthy environment and sustainable development are interdependant and the environment degradation is the result of the inequality of religion and social differences; this is why the two famous Economists namely Prof. B.Maire and J.Staldge had mentioned in their book 'Frontiers of Development' that there should be a co-operation between quality of development and environmental progress because this also reduces poverty by increasing employment opportunities. In present scenario the environment problems are directly related to food and water.

Economy Environment Nexus

The economy environment nexus can be identified in term of 5 major attributes which are:

1. Level of economic development (say per capita income)
2. Level of environment quality (say degree of deforestation).
3. Extend of poverty (say head count ratio, Sen Index)
4. Extend of income inequality (say Gini coefficient).
5. Population pressure (say density of population)

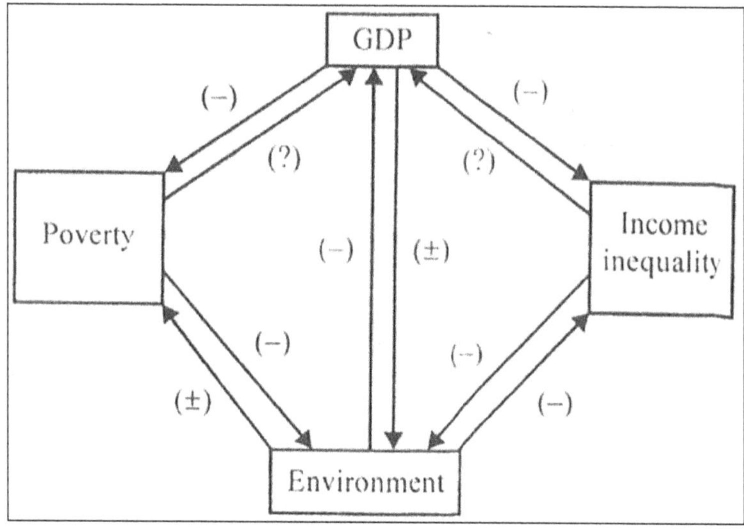

This figure shows the possible direct linkages between these economic and environment attributes. (+) sign indicates a direct positive relation whereas (–) negative sign is a negative one. Wherever? mark is put, it implies of not having any definite direction of relation. The empirical evidence however is quit mixed because of which no definite set of data can be shown as clear evidence. What kind of relation does one expect between them is a complex question. Theoretically value loaded and empirically is often difficult to establish.

Total Economic value of Environment

The Economic value of the environment need to be specified in the background of the three important features of environmental good, namely: 1) irreversibility, 2) uncertainty and 3) uniqueness. The decision to use up environmental goods is generally an irreversibility decision as the environment cannot be regenerated in a factory according to our wish and will. For the environment's regenerative and assimilative capacity one has to rely on the speed of nature's process. As we draw more and more resources from the environment and dump, and move faster into the environment, the entropy increases. Thus, the decision to use up environmental goods/ services is irreversible. Uncertainties arise form our limited knowledge of the ecological balance. If we use up some environmental goods we really donot know what else we are likely to lose in future. It is equally true about is the economic cost of green revolution in India seeing the productivity of soil has gone down in most of the districts of Punjab.

Conclusion

Today global common wealth is an important topic and for this there must be a world wide global policies for applications of sustainable development and consequences of reduction in economic disparity. Only then there will be an equal and proportionate equilibrium between economic growth and environment.

References

1. Bhattacharya, N. Rabindra (2001). *Environmental Economics*. Oxford India Papersbacks.
2. Bharucha, Erach (2005). *Environmental Studies*, Universities Press.
3. Saxsena, Binod Bihari (1995). *Parayawaran Parishitiki Evom Swasthaya*. Madhya Pradesh Hindi Granth Akadami.
4. Rastogi, Ramesh (2004). *Educational Development and Economy*, Sahitaya Sadan.

Chapter 20

Water Crisis and Deterioration of their Characteristics: An Urgent Need of Water Management and Planning for Sustainable Urban Development

Amardip Singh, Poonam and Bindeshwari Prasad Singh

ABSTRACT

In Tarn Taran urban area, water requirement for various uses such as domestic, agriculture and industry is mostly met from the groundwater sources. The groundwater characteristic in various areas of town especially in industrial and residential areas along the Kasur Nallah and Muradpur drain is unfit for uptake. In comparison to deep water aquifer, shallow water is seriously affected. On the basis of visual observation, surface water characteristics *i.e.*, colour, odour, turbidity and taste of nallah and drain flowing through the town is strongly objectionable.

Groundwater characteristics of Tarn Taran urban area have become vulnerable day by day due to unplanned urbanization, industrialization, intensive agriculture and unscientific disposal of solid waste. Intensive cultivation of crops causes chemicals from fertilizers (e.g. nitrate) and pesticides to seep into the groundwater. Discharge of untreated sewage and industrial effluents containing hazardous waste are finally drain out into

Kasur nallah and Muradpur drain and low laying areas, contaminate the groundwater. Pollution also occurs when aquifers are recharged with irrigation water contaminated with agricultural chemicals and fertilizers. Intense competition among users such as agriculture, industry and domestic sectors is driving the groundwater table lower. The characteristic of groundwater is also getting severely affected from the leachate due to unscientific disposal of solid wastes.

An attempt has therefore, been made to study the magnitude of water pollution in the Tarn Taran urban area during the period February 2009 to December 2009. The present study revealed that, there is an urgent need of integrated water management in spatial planning to keep the balance between urban development and water system so as to attain a sustainable urban system. It may be possible by adopting various scientific control measures including installation of sewage treatment plant (STP), effluent treatment plant (ETP) and common effluent treatment plant (CETP), de-siltation of nallah and drains, conservation measures, education and awareness, participation of NGO's and media, statutory obligations etc.

Keywords: *Urbanization, Industrialization, Intensive agriculture, Solid waste, Water scarcity, Water characteristics, Planning and management, Sustainable urban development.*

Introduction

Extensive urbanization and intensive agriculture, urban land use change in many cities and towns of the country has resulted in irrepressible disturbances of hydrological systems via reclamation, alteration and pollution. Surface water systems possess at least three groups of functions in urban region i.e, natural ecological functions, social and economic functions and physical spatial functions. Proper utilization and protection of these functions are important for sustainable urban development. The greater impacts of environment quality pollution have drawn increasing attention of scientists to assess water environmental damage (Batty 1971; Asako 1980; Galuzzi 1996; Grigg 1997; Marans 2003; Trauth and Shin 2005; Uzzell and Moser 2006)[1–7]. The majority of surface water pollution comes from non-point sources which are generated from various land use types and land use density (Marsh and Jhon Grossa, 2005, p 282-290)[8]. Discharge of untreated sewage and industrial effluents containing hazardous waste are finally drain out into nallah, drain, sewerage system and low laying areas, contaminate the groundwater, thereby deteriorating the characteristics of fresh water resources. Many researchers have quantified point source pollution (Ghosh 1997; Abbasi *et al.*, 2002; Paliwal *et al.*, 2007)[9-11].

Different factors contributing to the quality of urban runoff have been studied and it has been concluded that surface water quality is indeed affected by urbanization and industrialization (Al-Kharabsheh, 1999; Karn and Harada, 2001; Wang, 2001; Marsalek *et al.*, 2002; Kelsey *et al.*, 2004; Jamwal *et al.*, 2008)[12-17]. The poor quality of surface waters can also be attributed to the runoff generated from both dry and wet weather (Wenwei *et al.*, 2003; Taebi and Droste, 2004; Petersen *et al.*, 2005; McLeod *et al.*, 2006)[18-21]. Pollution by agricultural run-offs has too main effects on the environment. Environment-related infections and parasitic diseases thrive where there is a lack of clean drinking water, sanitation and drainage, and where air quality is poor (Nunan and Satterthwaite, 1999)[22].

Groundwater characteristics of Tarn Taran urban area have become vulnerable day by day due to unplanned urbanization, industrialization, intensive agriculture and unscientific disposal of solid waste. In the urban area, water requirement for various uses such as domestic, agriculture and industry is mostly met from the groundwater sources. The groundwater characteristic in various areas of town especially in industrial and residential areas along the Kasur nallah and Muradpur drain is unfit for uptake. In comparison to deep water aquifer, shallow water is seriously affected.

On the basis of visual observation, surface water characteristics *i.e.*, colour, odour, turbidity and taste of nallah and drain flowing through the town is strongly objectionable. Intensive cultivation of crops causes chemicals from fertilizers (*e.g.* nitrate) and pesticides to seep into the groundwater. Discharge of untreated sewage and industrial effluents containing hazardous waste are finally drain out into Kasur nallah and Muradpur drain and low laying areas, contaminate the groundwater. Intense competition among users such as agriculture, industry and domestic sectors is driving the groundwater table lower. The characteristic of groundwater is also getting severely affected from the leachate due to unscientific disposal of solid wastes.

An attempt has therefore, been made to study the magnitude of water pollution in the Tarn Taran urban area during the period February 2009 to December 2009. The present study revealed that, there is an urgent need of integrated water management in spatial planning to keep the balance between urban development and water system so as to attain a sustainable urban system. It may be possible by adopting various scientific control measures including installation of sewage treatment plant (STP), effluent treatment plant (ETP) and common effluent treatment plant (CETP), de-siltation of nallah and drains, conservation measures, education and awareness, participation of NGO's and media, statutory obligations etc.

Study Area

The Tarn Taran is a recently formed district of Punjab falling in Jalandhar Division. Tarn Taran district has been carved out of Amritsar district of which it was a part till the year 2006. Tarn Taran Local Planning Area (LPA) extends to 113.80 sq. km and has a population of 115936 persons, according to census 2001. The exact location of Tarn Taran is at 31° 27' 0" North and 74° 55' 31" East. The terrain of Tarn Taran can be put under three categories: the upland plain, bluff along the Beas and floodplain of Satluj.

The climate is generally dry except in the south-west monsoon season, a hot summer and bracing winter. The year may be divided in four seasons. The cold season is from November to March when minimum temperature reaches 4°C. The period from April to June is the hot season when maximum temperature reaches 40°C. The southwest monsoon season is from about the beginning of July to the first week of September. Tarn Taran received an annual average rainfall 389 mm. in 2006.

Materials and Methods

This study is based on physical survey of the water polluted areas of Tarn Taran town and its vicinity. Physical survey includes the visual observations of the polluted nallah and drain flowing through the town. Beside this, an intensive study of the characteristics of surface water of the river Beas and Sutlej for reference flowing near of the Tarn Taran urban area was also made. In addition, location and spatial distribution of water bodies, health affected zones, odour zone and mosquito zones along the Kasur nallah and Muradpur drain were studied intensively.

Results and Discussion

Water Bodies

The major natural features in vicinity of the Tarn Taran urban area are the river Beas and Sutluj. Besides, other water bodies within the LPA of Tarn Tarn included branch, distributories, minor, sarovar and ponds are also contributing a major role in recharging the groundwater source. Water from these sources is fresh and primarily used in to fulfill the irrigational requirements of the farmers. For drinking and other purposes, groundwater resource is used in and around the entire urban area. Kasur nallah and Muradpur drain flowing through the Tarn Taran urban area and of its vicinity, carries excessive amount of untreated sewage and industrial effluents.

Location and Spatial Distribution

The two important perennial rivers in vicinity of Tarn Tarn urban area are the river Beas flowing from north east to south west at 22.0 km from the town and river Sutluj join with river Beas near Harike Pattan flowing from east to west direction at 33.5 km from the town. Besides these, in the southern part of Tarn Taran LPA (Local Planning Area), Kasur Branch is flowing from east to west and feed the water supply to distributary and minor. Rasulpur distributory is originated from Kasur branch and Jodhpur minor is originated from Rasulpur distributory. Dobarji distributory is also flowing in the LPA of Tarn Taran.

Within the Tarn Taran LPA, Kasur nallah and Muradpur drain is flowing. Kasur nallah flows from east to west and towards west it leaves the LPA boundary. Muradpur drain flows from south east to north west direction. It enters into the LPA boundary from south east region of the Tarn Taran and after travelling certain distances it merges into Kasur nallah near Kazikot village.

Surface Water Characteristics

River

All parameters of surface water characteristics are within the permissible limits prescribed by BIS (ISI) except BOD and total coliforms. BOD and total coliform level for river Beas is higher than the standards prescribed by BIS for tolerant limit of Class A category (surface water). The detailed water characteristic parameters are summarized in Table 20.1.

Table 20.1: Status of Water Characteristics of River Beas and Sutlej

Sl.No.	*Parameter*	*Beas*	*Sutlej*
1.	Temperature	°C16	16
2.	pH	7.8	7.7
3.	Conductivity (mmho)	342	378
4.	Nitrogen (NO_2 + NO_3)	1.4	1.0
5.	DO (mg/l)	7.8	7.7
6.	BOD (mg/l)	4.2	1.8
7.	COD (mg/l)	14.4	6.4
8.	Cl^- (mg/l)	23.0	20
9.	SO_4	16	14
10.	Na	14.6	4.2
11.	Fecal Coliform	500	170
12.	Turbidity (NTU)	24	22
13.	Total Coliform	5000	500
14.	TDS	302	340

Source: PPCB, Dec 2000.

Kasur Nallah and Muradpur Drain

On the basis of visual observations, colour, odour, turbidity and smell, the characteristics of the effluents flowing in nallah and drain is strongly objectionable. It carries excessive amount of untreated industrial and sewage effluents of entire urban area, waste materials, decomposed weeds and shrubs and also having excessive of silts and mud. The level of pollution in nallah and drain is extremely high.

Surface Water Pollution

Unplanned urbanization, industrialization, intensive agriculture and unscientific disposal of solid waste has affected both surface and groundwater sources. Kasur nallah and Muradpur drain flows in the town being the major recipient of town's untreated sewage and industrial effluents, dumping of solid waste, dumping of ash from burnt rice husk etc. cause contamination of undergroundwater, unhygienic, foul smell and dampness. The level of pollution in these nallah and drain is extremely high.

Along the drain many small, medium and few large scale industrial units are existing and drain became the ultimate recipient of their untreated effluents. Due to unawareness about the health impact of hazardous chemical effluents which is directly discharged into the drain, illiterate cattle grazer left free their cattle's along the nallah and drain for drinking water and also for bathing. People residing near and along the drain and nallah claimed that, they are suffering from diseases like gastroenteritis, jaundice, diarrhea/ dysentery, malaria etc.

Health Affected Zone

Population residing near and along the nallah and drain has been subjected to water borne diseases like gastroenteritis, dysentery, jaundice and also malaria. Health affected zone along the nallah and drain in terms of percentage of area affected are summarized in Tables 20.2 and 20.3 respectively.

Table 20.2: Kasur Nallah Health Affected Zone

Zones (Distance from the Source)	*Per cent of Area Affected*
High effected zone upto 1000 meters	17.61

Note: Percentage of affected area is based on the total area of the LPA.

Source: Field surveys.

Table 20.3: Muradpur Drain Health Affected Zone

Zones (Distance from the Source)	*Per cent of Area Affected*
High effected zone upto 1000 meters	14.36

Note: Percentage of affected area is based on the total area of the LPA.

Source: Field surveys.

Odour Zone

For Kasur nallah, it has been found that, 350 m. wide belt either side of the nallah is affected by strong nuisance and bad odour. Highest affected area falls within 75 m. belt of nallah. 76 to 280 m. belt is moderately affected and between 281 to 350 m. of belt is least impacted area.

Table 20.4: Kasur Nallah Odour Zone

Odour Zones	*Distance from the Source*	*Per cent of Area Affected*
High odour	75 meter	1.16
Moderate odour	76 to 280 meter	3.19
Low odour	281 to 350 meter	1.07

Note: Percentage of area affected is based on total area of LPA. Odour zone is calculated only for those areas which are affected by Nallah (downstream from village Malia to end of the LPA).

Source: Field surveys.

For Muradpur drain, it has been found that a 325 wide belt either side of the drain is affected by strong nuisance of bad odour. Highest affected area falls within 60 m belt of drain. 61 to 260 m belt is moderately affected and between 261-325 m of belt is least impacted area (Table 20.5).

Table 20.5: Muradpur Drain Odour Zone

Odour Zones	*Distance from the Source*	*Per cent of Area Affected*
High odour	60 meter	0.66
Moderate odour	61 to 260 meter	2.17
Low odour	261 to 325 meter	0.70

Note: Percentage of area is based on the total area of the LPA. Odour zone is calculated only for those areas which are affected by drain (downstream from Sange to end of the LPA).

Source: Field surveys.

Mosquito Zone

Occurrence of malaria has been found to be rampant within a belt of 1000 m. either side of nallah and drain. The details of area affected due to mosquitoes are given in Tables 20.6 and 20.7. Worst sufferers have been found to live in 150 m. belt away from the source. Next 151 to 500 m. belt is moderately affected. Least affected area falls in between 500 to 1000 m.

Table 20.6: Kasur Nallah Mosquito Zones

Mosquito Zones	*Distance from the Source*	*Per cent of Area Affected*
High affected	upto 150 meters	2.63
Moderate affected	151-500 meters	6.14
Low affected	501-1000 meters	8.77

Note: Percentage of area affected is based on the total area of the LPA.

Source: Field survey.

Table 20.7: Muradpur Drain Mosquito Zones

Mosquito Zones	*Distance from the Source*	*Per cent of Area Affected*
High affected upto	150 meters	2.14
Moderate affected	151-500 meters	5.00
Low affected	501-1000 meters	7.16

Note: Percentage of area affected is based on the total area and population of the LPA.

Source: Field survey.

Groundwater Pollution

The degradation of the characteristics of groundwater caused due to excessive pollution level of Kasur nallah and Muradpur drain. The seepage of polluted water from both nallah and drain and the industrial waste has led to the pollution of the groundwater sources. Based on this fact, groundwater characteristic within the town is undesirable. Groundwater in most of the area which is passing along the nallah and drain has become unfit for drinking. The colour, odour, taste and presence of fine suspended particles are the cause of objection for their potable use. The colour of water is yellowish, odour is objectionable and suspended particles can be visualized by necked eyes. In comparison to deep water aquifer, shallow water is seriously affected. Majority of the residents of Tarn Taran town residing along nallah and drain and that of other adjoining villages are forced to consume contaminated vegetables and drink unsafe water, thus exposing themselves to the risk of water-borne diseases.

Major issues emerging from the groundwater pollution have been listed below:

- ✰ Persons residing in abadies in close proximity to Kasur nallah and Muradpur drain have been found to be exposed to water borne diseases due to polluted groundwater.
- ✰ Considerable level of groundwater pollution have been found to exist upto a depth of 100 ft. along the 1000 meter belt on either side of Kasur nallah and Muradpur drain.
- ✰ Pollution of soil and groundwater has also been caused by dumping of industrial wastes into the open ground leading to stagnation and the generation of the leachate.
- ✰ Use of polluted groundwater for agricultural purposes has also led to the degradation of soil and presence of hazardous chemicals into soil and vegetable crops.

Control Measures

Considering the above critical situation prevailing in the study area, there is an urgent need to control water pollution for sustainable urban development by adopting various scientific control measures including installation of sewage treatment plant (STP), effluent treatment plant (ETP), de-siltation of nallah and drains, conservation measures, education and awareness, participation of NGO's and media, statutory obligations etc.

Installtion of STP, ETP and CETP

Sewage treatment plant (STP) is the most efficient system in which sewage water is treated by physical, chemical and biological means and after treatment water becomes free from water pollutants. Water pollutant such as BOD, COD, TSS, etc. levels reduces at permissible level and enhances the concentration of DO level. Effluent Treatment Plant (ETP) and Common Effluent Treatment Plant (CETP) is also an effective system for treatment of industrial effluents. It also reduces the excessive concentration of water pollutants to permissible level. At present town has lack of STP and also has insufficient no. of ETP and CETP in the industrial area. Hence, it is an urgent need for installation of STP, ETP and CETP in and around the town and industrial area.

Conservation Measures

There may be various methods to conserve water bodies and polluted nallah and drain by adopting the following control measures *i.e.*, measures for rejuvenation of Kasur nallah and Muradpur drain, conservation of ponds, restoration plan, rain water harvesting, solid waste management etc.

Measures for Rejuvenation of Kasur nallah and Muradpur drain

Following measures to be taken to rejuvenate the Kasur nallah and Muradpur drain *i.e.*:

- ✰ Banning of discharge of untreated sewage and industrial effluents into nallah and drain.
- ✰ Demarcation of entire length of nallah and drain and removal of encroachments.
- ✰ Periodical desilting of nallah and drain should be ensured.
- ✰ The entire length of nallah and drain especially within the town limits should be paved.
- ✰ Conversion of whole nallah and drain into a green belt acting as a bio point runoff.
- ✰ Dustbins on selective sites along nallah and drain be provided for solid waste collection.
- ✰ The solid waste dumped into nallah and drain should be removed and shifted to designated land fill sites of Municipality.
- ✰ No solid waste or cow dung should be allowed to be dumped in and along nallah and drain.
- ✰ New land fill sites should be identified for the disposal of solid waste of all categories.
- ✰ All the industrial units in and around Tarn Taran town should be directed to set up ETP/CETP individually or collectively to achieve zero liquid discharge.
- ✰ Public awareness campaign should be carried out and the public should be awakened not to spoil the drainage system by throwing domestic/industrial waste into nallah and drain.

Conservation of Pond and Restoration Plan

Ponds in the Tarn Traan urban area and of its vicinity are in varying degrees of environmental degradation. The degradation is due to encroachments, eutrophication and continuous siltation. The pond restoration plan and action could be categorized broadly as elaborated in the following management actions.

Source Control

- ☆ Treatment of watershed or catchment of ponds which are not only brings in substantial improvements in the pond environment (reduction of silt, control of chemicals and nutrients) but also results in overall development of the community living in catchment.
- ☆ Soil conservation measures, bank/slope erosion control measures, afforestation, drainage improvements, control of sewage wastes, sewage interceptions and diversions and participation of people in watershed management measures have been widely adopted as effective management tools in all the pond restoration projects.

Pond Treatment

Following are the several palliative measures under taken to remove eutrophication and improve quality of pond water.

- ☆ Dredging and de-silting
- ☆ De-weeding/hyacinth control or removal (biological, chemical, mechanical and manual measures, bio-composting)
- ☆ Bio-remediation (clean up with bio-products - natural bacteria breakdown, and aerators to churn the pond/lakes)
- ☆ Introduction of composite fish culture/larvivorous fish species to control mosquitoes
- ☆ Harvesting of aquatic weeds

Rainwater Harvesting

Now a days town planners and civic authority in many cities and towns of India are introducing bylaws making rainwater harvesting compulsory in all new structures. No water or sewage connection would be given if a new building did not have provisions for rainwater harvesting. Such rules should also be implemented in Tarn Taran town to ensure a rise in the groundwater level.

Environmental Education and Awareness

Environmental education awareness in general, which includes wetlands as well, is included in curricula at different levels of education. Special modules relating to wetlands have been developed at Wildlife Institute of India to generate awareness about the values and functions of wetlands and the need for their conservation to wider group of participants representing government, non-government and private entrepreneurs.

Peoples's Participation

This very effective management method is becoming increasingly popular in conserving water bodies in urban areas. Non-Governmental Organizations have acted as catalysts. In major urban centers, people have organized themselves. They have also moved the Judiciary (the Supreme Court and the High Courts) through Public Interest Litigations (PILs) seeking directives of the courts to restore urban water bodies including ponds, drains, nallah canal etc. Information Centers-cum-watch towers for mass awareness and promoting public participation in the water bodies conservation programme in this regard is essential. In this direction the Ministry of Environment and Forests (MOEF) has recognized that the elaborate process of assessment of social, economic and ecological aspects of water bodies resources through community participation can help to formulate a comprehensive management plan, which is ecologically viable and socially acceptable.

Statutary Obligations

In India, various legislations related to water pollution and uses have been formulated by the Ministry of Environmental and Forest such as Water (Prevention and control of pollution) Act 1974, the Water Cess Act 1977, The Environment (Protection) Act 1986, Environment (Sitting for Industrial Projects) Rules 1999,

Water (Prevention and Control of Pollution) Act 1974

The basic objective of the act is to maintain and restore the wholesomeness of national aquatic resources by prevention and control of pollution. The act is comprehensive in its coverage, applying to streams, inland waters, subterranean waters and sea or tidal waters. Standards for the discharge of effluent or quality of the receiving waters are not specified in act, but it enables state boards and central board to prescribe these standards. The act prohibits disposal of polluting matter in streams, wells and sewers or on land in excess of the standards prescribes by the state boards. It is must to obtain consent from the state pollution control boards before establishing any industry, operation process, any treatment and disposal system or any extension or addition to such a system which might result in the discharge of sewage or trade effluent into stream, well or sewer or onto land. The water act establishes a central and state pollution control boards.

Water Cess Act 1977

To strengthen the pollution control boards by providing money for equipments and technical personnel and to promote water conservation by recycling, Parliament adopted this act. The act empowers the central government to impose a cess on water consumed by industries listed in schedule I of the act.

Environment (Protection) Act 1986

The purpose of this act is to implement the decisions of the United Nations Conference on the Human Environment of 1972, so far as they relate to the protection and improvement of the human environment and prevention of hazards to human beings, other living creatures, plant and property. The act provides a framework for central government coordination of the activities of various central and state authorities established under previous laws. The act empowers the centre to take all such measures as it deems necessary or expedite for the purpose of protecting and improving the quality of the environment and preventing, controlling and abating environmental pollution. The act explicitly prohibits discharge of pollutants in excess of prescribed standards. Persons responsible for discharge of pollutants in excess of standards must prevent or mitigate the pollution and must report the discharge to government authorities. The act provides for severe penalties such as imprisonment and monetary fine. The act empowers the central government to establish standards for the quality of the environment in its various aspects for different areas. In exercise of powers conferred by the act, Water Quality Assessment Authority at central level and state level Water Quality Review Committee have been constituted.

Environment (Siting for Industrial Projects) Rules 1999

Prohibition for Setting up of Certain Industries

No new unit of the industries listed in Annexure-I shall be allowed to be set up in the following areas:

- ☆ The entire area within the municipal limits of all Municipal Corporations, Municipal Councils and Nagar Panchayats (by whatever name these are known in each state) and a 25 km belt around the cities having population of more than 1 million.
- ☆ 7 km belt around the periphery of the wetlands listed in Annexure-II; and
- ☆ 25 km belt around the periphery of National Parks, Sanctuaries and core zones of Biosphere Reserves,
- ☆ 0.5 km wide strip on either side of national highways and rail lines.

Establishment of New Units with Certain Conditions

- ☆ Establishment of new units of the industries listed in Annexure-I shall be allowed in 7 km to 25 km belt around the periphery of the wetlands listed in Annexure-II only after careful assessment of their adverse ecological and environmental impacts.
- ☆ Establishment of New Units around Archaeological Monuments.
- ☆ New units of industries listed in Annexure-III shall not be allowed to be set up within 7 km periphery of the important archaeological monuments listed in Annexure-IV.

Conclusion

The present study revealed that, groundwater characteristics of Tarn Taran urban area have become vulnerable day by day due to unplanned urbanization, industrialization, intensive agriculture and unscientific disposal of solid waste. Intensive cultivation of crops causes chemicals from fertilizers (*e.g.* nitrate) and pesticides to seep into the groundwater. Discharge of untreated sewage and industrial effluents containing hazardous waste, finally drain out into Kasur nallah and Muradpur drain and low laying areas, contaminate the groundwater. Pollution also occurs when aquifers are recharged with irrigation water contaminated with agricultural chemicals and fertilizers. Intense competition among users such as agriculture, industry, and domestic sectors is driving the groundwater table lower. The characteristic of groundwater is also getting severely affected from the leachate due to unscientific disposal of solid wastes.

Considering the above critical situation prevailing in the town and also of its vicinity within the LPA, there is an urgent need to control the water pollution for sustainable urban development by adopting various scientific control measures including installation of sewage treatment plant (STP), effluent treatment plant (ETP), common effluent treatment plant (CETP), de-siltation of nallah and drain, conservation measures, education and awareness, participation of NGO's and media, statutory obligations etc.

Acknowledgement

This study was based on physical survey of the water polluted zones of Tarn Taran urban area and its vicinity carried out independently by the author (project was funded by PUDA) and also the corresponding author was the former employee of the SAI Consulting Engineers Pvt. Ltd.

References

1. Batty, M. (1971). Modelling cities as dynamic systems. *Nature (London)*, 231: 425–428.
2. Asako, K. (1980). "Growth and environmental pollution under the maximum.
3. Galuzzi, M.R. (1996). Integrating drainage, water quality, wetlands, and habitat in a planned community development. *J. Urban Plann. Dev.*, 122(3): 101–108.
4. Grigg, N.S. (1997). Systemic analysis of urban water supply and growth management. *J. Urban Plann. Dev.*, 123(2): 23–33.
5. Marans, R. (2003). Understanding environmental quality through quality of life studies: The 2001 DAS and its use of subjective and objective indicators. *Landsc. Urban Plann.*, 65: 73–83.

6. Trauth, K.M. and Shin, Y.-S. (2005). Implementation of the EPA's water quality trading policy for storm water management and smart growth. *J. Urban Plann. Dev.*, 131(4): 258–269.

7. Uzzell, D.L. and Moser, G. (2006). Environment and quality of life. *European Review of Applied Psychology*, 56(1): 1–4.

8. Marsh, W.M. and John Grossa, J. (2005). *Environmental Geography: Science, Land Use and Earth Systems*, 3rd edn. John Wiley and Sons, Inc., USA.

9. Ghosh, N.C. (1997). Uncertainty analysis of CBOD and DO profiles of Kali river (India) water quality using QUAL2E–UNCAS. *Indian Journal of Environmental Health*, 39(1): 8–19.

10. Abbasi, S.A., Khan, F.I., Sentilvelan, K. and Shabudeen, A. (2002). Modeling of buckingham canal water quality. *Indian Journal of Environmental Health*, 44(4): 290–297.

11. Paliwal, R., Sharma, P. and Kansal, A. (2007). Water quality modelling of the river Yamuna (India) using QUAL2E–UNCAS. *Journal of Environmental Management*, 83(2): 131–144.

12. Al-Kharabsheh, A.A. (1999). Influence of urbanization on water quality at Wadi Kufranja Basin (Jordan). *Journal of Arid Environments*, 43: 79–89.

13. Karn, K.S. and Harada, H. (2001). Surface water pollution in three urban territories of Nepal, India, and Bangladesh. *Environmental Management*, 28: 483–496.

14. Wang, X. (2001). Integrating water-quality management and land-use planning in a watershed context. *Journal of Environmental Management*, 61: 25–36.

15. Marsalek, J., Diamond, M., Kok, S. and Watt, E.W. (2002). *Urban Runoff*. The National Water Research Institute. <http: // www.nwri.ca/threatsfull/ch11–1–e.html>.

16. Kelsey, H., Porter, E.D., Scott, G., Neet, M., White, D. (2004). Using geographic information systems and regression analysis to evaluate relationships between landuse and fecal coliform bacterial pollution. *Journal of Session*, 1.710

17. Jamwal, P., Mittal, A.K. and Mouchel, J.M. (2008). Point and non-point microbial source pollution: A case study of Delhi. *Physics and Chemistry of the Earth*. doi: 10.1016/ j.pce.2008.09.005a

18. Wenwei, R., Zhong, Y., Meligrana, J., Anderson, B., Watt, W.E., Chen, J. and Leung, Hok-Lin (2003). Urbanization, landuse and water quality in Shanghai 1947–1996. *Environment International,* 29: 649–659.

19. Taebi, A. and Droste, L.R. (2004). Pollution loads in urban runoff and sanitary wastewater. *Science of the Total Environment,* 327: 175–184.

20. Petersen, M.T., Rifai, S.H., Suarez, P.M. and Stein, A.R. (2005). Bacteria loads from point and non-point sources in an urban watershed. *Journal of Environmental Engineering,* 131: 1414–1425.

21. McLeod, M.S., Kells, A.J., ASCE, M. and Putz, J.G. (2006). Urban runoff quality characterization and load estimation in Saskatoon, Canada. *Journal of Environmental Engineering,* 132: 1470–1481.

22. Nunan and Satterthwaite, (1999).

Chapter 21

Water Pollution Affecting the Agriculture

Chanda Kumari and R.K. Singh

ABSTRACT

Agriculture is a dominant component of the global economy. While mechanization of farnung in many countries has resulted in a dramatic fall in the proportion of population working in agriculture the pressure to produce enough food has had worldwide impact on agricultural practice. This pressure has resulted in expansion into marginal lands and is usually associated with subsistence farming. Foods required have required expansion of irrigation to achieve and sustain. Agro based industry and level of crops production are widely affected due to water pollution.

In India nearly about 35 per cent of population are facing low COD content with 0.003 ppm water with high impurity.The rate of population drive and contaminated water in agriculture are making more vital disease and mind, heart problem in our day to day activities. The mortality rate is rising due to inadequate food supplement and agriculture processing system is contaminated with high water impurity.

Introduction

Sustainable agriculture is one of the greatest challenges. Sustainability implies that agriculture not only secure a sustained food supply but that its environmental, socio-economic and human health impacts are recognized and accounted for their natural development plans.

Action items for Agriculture in the field of water quality as:

- ☆ Establishments and operation of cost –effective water quality monitoring system for agriculture water uses
- ☆ Prevention of adverse effects of agriculture activities in water quality for other social and economic activities and on wetland inter alia through optional use of on farm inputs and the minimization of the use of external inputs in agricultural activities.
- ☆ Establishments of biological,physical and chemical water quality criteria for agriculture water uses and for marine and reverie ecosystems.
- ☆ Prevention of soil run off and sedimentation.
- ☆ Proper disposal of sewage from human settlements and of manure produced by intensive livestock breeding.
- ☆ Minimization of adverse effects from agriculture chemicals by use of integrated post management.

Education if communities about the pollution impacts of the of fertilizer and chemical or water quality and food safety.

Water Quality as a Global Issue

The principle of environmental and public health dimensions of the global freshwater quality problem are highlighted below.

- ☆ Five million people die annually from water borne disease.
- ☆ Ecosystem disfunction and loss of bio diversity.
- ☆ Contamination of marine ecosystem from land based activities.
- ☆ Contamination of groundwater resources.
- ☆ Global contamination by persistent pollutants.

The Crisis is Predicted to have the Following Global Dimension

- ☆ Decline in sustainable food resources(*e.g.*fresh water and coastal fisheries) due to pollution.

- ☆ Cumulative effect of poor water resources management decisions because of inadequate water quality data in many countries.
- ☆ Many countries have no longer pollution management leading to higher levels of aquatic pollution.
- ☆ Escalating cost of remediation and potential loss of "credit worthiness".

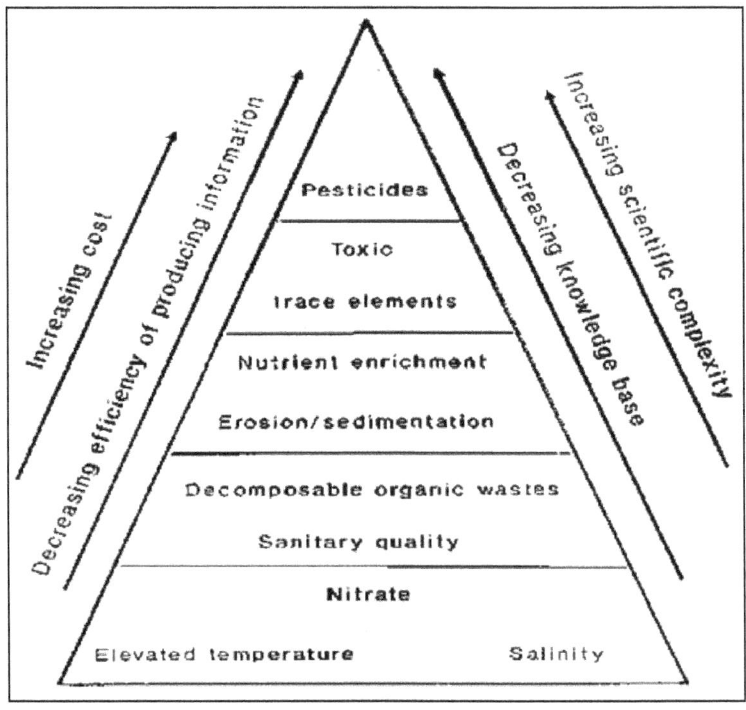

Figure 21.1: Hierachial Complexity of Agriculturally Related Water Quality Problems

Irrigation Impacts on Surface Water Quality

United Nations' predictions of global population increase to the year 2025 require an expansion of food production of about 40-45 per cent. Irrigation agriculture, which currently comprises 17 per cent of all agricultural land yet produces 36 per cent of the world's

Table 21.1: Per cent of Assessed River Length and Lake Area Impacted

Source of Pollution	*Rivers (per cent)*	*Lakes (per cent)*	*Nature of Pollutant*	*Rivers (per cent)*	*Lakes (per cent)*
Agriculture	72	56	Siltation (sediment)	45	22
Municipal point sources	15	21	Nutrients	37	40
Urban runoff/storm sewers	11	24	Pathogens	27	
Resource extraction	11		Pesticides	26	
Industrial point sources	7		Organic enrichment DO	24	24
Silviculture	7		Metals	19	47
Hydrologic/habitat modification	7	23	Priority organic chemicals		20
On-site wastewater disposal		16			
Flow modification		13			

food, will be an essential component of any strategy to increase the global food supply. Currently 75 per cent of irrigated land is located in developing countries; by the year 2000 it is estimated that 90 per cent will be in developing countries.

In addition to problems of waterlogging, desertification, salinization, erosion, etc., that affect irrigated areas, the problem of downstream degradation of water quality by salts, agrochemicals and toxic leachates is a serious environmental problem. "It is of relatively recent recognition that salinization of water resources is a major and widespread phenomenon of possibly even greater concern to the sustainability of irrigation than is that of the salinization of soils, per se. Indeed, only in the past few years has it become apparent that trace toxic constituents, such as Se, Mo and As in agricultural drainage waters may cause pollution problems that threaten the continuation of irrigation in some projects" (Letey *et al.*, cited in Rhoades, 1993).

Public Health Impacts

Polluted water is a major cause of human disease, misery and death. According to the world health organization (WHO)as many as 4 million children die every year as a result of diarrhoea caused by water borne infections.Agricultural pollution is both a direct and indirect cause for the impacts on human health. WHO reports that nitrogen level in groundwater has grown in many parts of the world as a result of intensification of farming practices.

Conclusion

The environmental impacts on water resources [in general] caused by agricultural activities cannot be dissociated from the agricultural impacts in production areas themselves. They require monitoring, and preventive measures should always be systemically integrated.

It is necessary to develop and implement water resource monitoring systems with a prior definition of indicators, parameters, tolerance limits, frequency and sampling points, combining this information with quantity data.

Ultimately, any strategy to reduce agricultural impacts on water quality will only be successful if it is implemented at the farm level.

Therefore, implementation of control measures at the farm level will only be successful and sustainable if the farmer can determine that it is in his economic interest to undertake such measures. Therefore, the economic benefits from such factors as implementation of erosion control measures to maintain soil fertility, capital costs associated with improved manure handling and distribution, etc., must be clearly seen to be offset by reduced.

References

1. Abbott, M.B., Bathurst, J.C., Cunge, J.A., O'Connell, P.E. and Rasmussen, J. (1986). An introduction to the European Hydrological System – Système Hydrologique Européen "SHE". *J. Hydrol.*, 87: 45–77.
2. Ackefors, H. and Enell, M. (1992). Pollution loads derived from aquaculture: Land-based and water-based systems. In: *Workshop on Fish Farm Effluents and their Control in EC Countries.* Published by the Department of Fishery Biology, Institute for Marine Science at the Christian-Albrechts-University of Kiel, Germany, pp. 3–4.
3. Andreoli, C.V. (1993). The influence of agriculture on water quality. In: *Prevention of Water Pollution by Agriculture and Related Activities.* Proceedings of the FAO Expert Consultation, Santiago, Chile, 20–23 Oct. 1992. Water Report 1. FAO, Rome pp. 53–65.
4. Appelgren, B.G. (1994). *Agricultural and Environmental Legislation – Lithuania,* Technical Report. FAO–LEG: TCP/LIT/2352, Technical Cooperation Programme, FAO, Rome.
5. Avcievala, S. (1991). *The Nature of Water Pollution in Developing Countries.* Natural Resources Series No. 26. UNDTCD, United Nations, New York.
6. Bangay, G.E. (1976). *Livestock and Poultry Wastes in the Great Lakes Basin: Environmental Concerns and Management Issues.* Social Science Series No. 15. Environment Canada.
7. Barg, U.C. (1992). *Guidelines for the Promotion of Environmental Management of Coastal Aquaculture Development.* FAO Fisheries Technical Paper 328, FAO, Rome.
8. Beasley, D.B. and Huggins, L.F. (1981). *Answers Users Manual.* US–EPA 905/9–82–001. US Environmental Protection Agency, Washington DC.

9. Braune, E. and Looser, U. (1989). Cost impacts of sediments in South Africa rivers. In: *Sediment and the Environment,* (Eds.) R.F. Hadley and E.D. Ongley. IAHS Publication, Int. Assoc. Hydrol. Sci., Wallingford, UK, 184: 131–143.

10. Bushway, R.J., Perkins, B., Savage, S.A., Lekousi.

Chapter 22

Water Pollution: Causes, Effects and Solutions

Veena Kumari, Mahesh Kumar Singh, Bina Pathak and Gopal Pal

ABSTRACT

Water pollution has emerged as one of the gravest environmental threats in India. Its biggest sources are city sewage and industrial waste that are discharged untreated into the rivers. Despite the best efforts of the government, only about 10 per cent of the wastewater that is generated in the cities is treated and the rest is discharged into the rivers.

The entry of toxic substances into water bodies like lakes, streams and rivers leads to deterioration in the quality of water and severely affects the aquatic ecosystems. Due to this, even the groundwater gets contaminated. All these have a devastating effect on all living creatures that exist near the polluted water bodies. Urgent steps are needed to be taken by the Indian government on the water pollution management front and the flawed policies need to be amended in order to obtain concrete results.

Introduction

Water pollution is a reality of human existence. Activities like agriculture and industrial production generate water pollution apart from the biological waste. In India, every year, approximately 50,000 million litres of wastewater, both industrial and domestic, is generated in urban areas. If the data of rural areas is also taken into account, the overall figure will be much higher. The materials that constitute industrial waste include highly harmful substances like salts, chemicals, grease, oils, paints, iron, cadmium, lead, arsenic, zinc, tin, etc. In some cases even radio-active materials are discharged into the rivers bodies by some companies, who for the sake of saving money on water treatment, throw all the norms to the winds.

All efforts by the government to put a check on wastewater management have failed as the treatment systems require high capital investment for installation and also high cost is incurred on operational maintenance. This is a sore point not only for the farmers but also for the factory owners as the high cost of treating industrial wastewater affects their bottom-line. The cost of establishing and running a wastewater treatment plant in a factory can be as high as 20 percent of the total expenditure. Hence we see a situation where, despite the presence of government norms, effluents continue to flow into the river bodies untreated.

On the other hand, the government of India is spending millions of rupees every year on water pollution control. According to rough estimates, Indian government has spent nearly 20,000 crore rupees till now on various schemes in India, like the Ganga Action Plan and Yamuna Action Plan, to control water pollution in rivers. But no positive results have been achieved as yet. The government should realise that all efforts to get the river-bodies free from water pollution will fail unless the process of untreated industrial and other wastewater getting into the water bodies is not stopped.

Hence the government should, instead of spending money on pollution control schemes, divert its resources to encourage wastewater treatment in agriculture and industrial sector. The money spent on pollution control should be spent on giving subsidies to the industries which generate wastewater and on strict monitoring of their adherence to the norms. Research should be promoted in areas like nanotechnology to find out ways and means to build cheaper wastewater management plants. Here also, the approach should be to re-use the treated water for agriculture instead of letting it go into the rivers and streams.

It should not be forgotten that only 0.3 per cent of total water available on this planet is fit for consumption for human beings, animals and plants. The remaining 99.7 per cent is present either as sea water or as glaciers on the mountains. Hence ignoring the issue of water pollution any longer would mean inviting serious complicacies in the future.

Now before we go through the causes,effects and solutions for the recent challenges faced by water pollution we should have a look at the following questions which would be sufficient to set a basic idea of water pollution into our minds.

Some Questions Related to Water Pollution

What is Water Pollution?

Water pollution is any chemical, physical or biological change in the quality of water that has a harmful effect on any living thing that drinks or uses or lives (in) it. When humans drink polluted water it often has serious effects on their health. Water pollution can also make water unsuited for the desired use.

What are the Major Water Pollutants?

There are several classes of water pollutants. The first are disease-causing agents. These are bacteria, viruses, protozoa and parasitic worms that enter sewage systems and untreated waste.

A second category of water pollutants is oxygen-demanding wastes; wastes that can be decomposed by oxygen-requiring bacteria. When large populations of decomposing bacteria are converting these wastes it can deplete oxygen levels in the water. This causes other organisms in the water, such as fish, to die.

A third class of water pollutants is water-soluble inorganic pollutants, such as acids, salts and toxic metals. Large quantities of these compounds will make water unfit to drink and will cause the death of aquatic life.

Another class of water pollutants are nutrients; they are water-soluble nitrates and phosphates that cause excessive growth of algae and other water plants, which deplete the water's oxygen supply. This kills fish and, when found in drinking water, can kill young children.

Water can also be polluted by a number of organic compounds such as oil, plastics and pesticides, which are harmful to humans and all plants and animals in the water.

A very dangerous category is suspended sediment, because it causes depletion in the water's light absorption and the particles spread dangerous compounds such as pesticides through the water.

Finally, water-soluble radioactive compounds can cause cancer, birth defects and genetic damage and are thus very dangerous water pollutants.

Where Does Water Pollution Come From?

Water pollution is usually caused by human activities. Different human sources add to the pollution of water. There are two sorts of sources, point and nonpoint sources. Point sources discharge pollutants at specific locations through pipelines or sewers into the surface water. Nonpoint sources are sources that cannot be traced to a single site of discharge.

Examples of point sources are: factories, sewage treatment plants, underground mines, oil wells, oil tankers and agriculture.

Examples of nonpoint sources are: acid deposition from the air, traffic, pollutants that are spread through rivers and pollutants that enter the water through groundwater.

Nonpoint pollution is hard to control because the perpetrators cannot be traced.

How Do we Detect Water Pollution?

Water pollution is detected in laboratories, where small samples of water are analysed for different contaminants. Living organisms such as fish can also be used for the detection of water pollution. Changes in their behaviour or growth show us, that the water they live in is polluted. Specific properties of these organisms can give information on the sort of pollution in their environment. Laboratories also use computer models to determine what dangers there can be in certain waters. They import the data they own on the water into the computer, and the computer then determines if the water has any impurities.

What is Heat Pollution, What Causes it and What are the Dangers?

In most manufacturing processes a lot of heat originates that must be released into the environment, because it is waste heat. The cheapest way to do this is to withdraw nearby surface water, pass it through the plant, and return the heated water to the body of surface water. The heat that is released in the water has negative effects on all life in the receiving surface water. This is the kind of pollution that is commonly known as heat pollution or thermal pollution.

The warmer water decreases the solubility of oxygen in the water and it also causes water organisms to breathe faster. Many water organisms will then die from oxygen shortages, or they become more susceptible to diseases.

What is Eutrophication, What Causes it and What are the Dangers?

Eutrophication means natural nutrient enrichment of streams and lakes. The enrichment is often increased by human activities, such as agriculture (manure addition). Over time, lakes then become eutrophic due to an increase in nutrients.

Eutrophication is mainly caused by an increase in nitrate and phosphate levels and has a negative influence on water life. This is because, due to the enrichment, water plants such as algae will grow extensively. As a result the water will absorb less light and certain aerobic bacteria will become more active. These bacteria deplete oxygen levels even further, so that only anaerobic bacteria can be active. This makes life in the water impossible for fish and other organisms.

What is Acid Rain and How Does it Develop?

Typical rainwater has a pH of about 5 to 6. This means that it is naturally a neutral, slightly acidic liquid. During precipitation rainwater dissolves gasses such as carbon dioxide and oxygen. The industry now emits great amounts of acidifying gasses, such as sulphuric oxides and carbon monoxide. These gasses also dissolve in rainwater. This causes a change in pH of the precipitation – the pH of rain will fall to a value of or below 4. When a substance has a pH of below 6.5, it is acid. The lower the pH, the more acid the substance is. That is why rain with a lower pH, due to dissolved industrial emissions, is called acid rain.

Why Does Water Sometimes Smell Like Rotten Eggs?

When water is enriched with nutrients, eventually anaerobic bacteria, which do not need oxygen to practice their functions, will become highly active. These bacteria produce certain gasses during their activities. One of these gases is hydrogen sulphide. This compounds smells like rotten eggs. When water smells like rotten eggs we can conclude that there is hydrogen present, due to a shortage of oxygen in the specific water.

Some Facts about Water Pollution in India

Out of India's 3,119 towns and cities, just 209 have partial treatment facilities, and only 8 have full wastewater treatment facilities (WHO 1992). 114 cities dump untreated sewage and partially cremated bodies directly into the Ganges River. Downstream, the untreated water is used for drinking, bathing, and washing. This situation is typical of many rivers in India as well as other developing countries.

Open defacation is widespread even in urban areas of India.Water resources have not therefore been linked to either domestic or international violent conflict as was previously anticipated by some observers. Possible exceptions include some communal violence related to distribution of water from the Kaveri River and political tensions surrounding actual and potential population displacements by dam projects, particularly on the Narmada River. Punjab is today another hotbed of pollution, for example, Buddha Nullah, a rivulet which run through Malwa region of Punjab, India, and after passing through highly populated Ludhiana district, before draining into Sutlej River, a tributary of the Indus river, is today an important case point in the recent studies, which suggest this as another Bhopal in making. A joint study by PGIMER and Punjab Pollution Control Board in 2008, revealed that in villages along the Nullah, calcium, magnesium, fluoride, mercury, β-endosulphan and heptachlor pesticide were more than permissible limit (MPL) in ground and tap waters. Plus the water had high concentration of COD and BOD (chemical and biochemical oxygen demand), ammonia, phosphate, chloride, chromium, arsenic and chlorpyrifos pesticide. The groundwater also contains nickel and selenium, while the tap water has high concentration of lead, nickel and cadmium.

The Mithi River, which flows through the city of Mumbai, is heavily polluted.

Sanitation

Most Indians depend on on-site sanitation facilities. Recently, access to on-site sanitation have increased in both rural and urban areas. In rural areas, total sanitation has been successful. In urban areas, a good practice is the Slum Sanitation Program in Mumbai that has provided access to sanitation for a quarter million slum dwellers.Sewerage, where available, is often in a bad state. In Delhi the sewerage network has lacked maintenance over the years and overflow of raw sewage in open drains is common, due to blockage, settlements and inadequate pumping capacities. The capacity of the 17 existing wastewater treatment plants in Delhi is adequate to cater a daily production of wastewater of less than 50 per cent of the drinking water produced. Of the 2.5 Billion people in the world that defecate openly, some 665 million live in India. This is of greater concern as 88 per cent of deaths from diarrhoea occur because of unsafe water, inadequate sanitation and poor hygiene.

Health Impact

The lack of adequate sanitation and safe water has significant negative health impacts including diarrhoea, referred to by travelers as the "Delhi Belly", and experienced by about 10 million visitors annually.While most visitors to India recover quickly and otherwise receive proper care, the World Health Organisation estimated that around 700,000 Indians die each year from diarrhoea. The dismal working conditions of sewer workers are another concern. A survey of the working conditions of sewage workers in Delhi showed that most of them suffer from chronic diseases, respiratory problems, skin disorders, allergies, headaches and eye infections

Water Supply and Water Resources

Depleting ground water table and deteriorating groundwater quality are threatening the sustainability of both urban and rural water supply in many parts of India. The supply of cities that depend on surface water is threatened by pollution, increasing water scarcity and conflicts among users. For example, Bangalore depends to a large extent on water pumped since 1974 from the Kaveri river, whose waters are disputed between the states of Karnataka and Tamil Nadu. As in other Indian cities, the response to water scarcity is to transfer more water over large distances at high costs. In the case of Bangalore, the 3,384 crore (US$751.2 million) Kaveri Stage IV project, Phase II, includes the supply of 500,000 cubic meter of water per day over a distance of 100 km, thus increasing the city's supply by two thirds

Causes of Water Pollution

Water Pollution: Point and Non-point Sources

There are many specific causes of water pollution, but before we list the toppers, it's important to understand two broad categories of water pollution:

Point Source

Occurs when harmful substances are emitted directly into a body of water.

Non-point Source

Delivers pollutants indirectly through transport or environmental change.

An example of a point source of water pollution is a pipe from an industrial facility discharging effluent directly into a river. An example of a nonpoint-source of water pollution is when fertilizer and chemical pesticides from a farm field is carried into a stream by rain (*i.e.* run-off).

Point-source pollution is usually monitored and regulated, at least in Western countries, though political factors may complicate how successful efforts are at true pollution control. Nonpoint sources are much more difficult to monitor and control, and today they account for the majority of contaminants in streams and lakes.

Now, on to the more specific categories of water pollution causes.

Cause 1: Pesticides

Pesticides that get applied to farm fields and roadsides–and homeowners' lawns–run off into local streams and rivers or drain down into groundwater, contaminating the fresh water that fish swim in and the water we humans drink. It's tempting to think this is mostly a farming problem, but on a square-foot basis, homeowners apply even more chemicals to their lawns than farmers do to their fields! Still, farming is a big contributor to this problem.

Cause 2: Fertilizers/Nutrient Pollution

Many causes of pollution, including sewage, manure, and chemical fertilizers, contain "nutrients" such as nitrates and phosphates. Deposition of atmospheric nitrogen (from nitrogen oxides) also causes nutrient-type water pollution.

In excess levels, nutrients over-stimulate the growth of aquatic plants and algae. Excessive growth of these types of organisms clogs our waterways and blocks light to deeper waters while the organisms are alive; when the organisms die, they use up dissolved oxygen as they decompose, causing oxygen-poor waters that support only diminished amounts of marine life. Such areas are commonly called dead zones.

Nutrient pollution is a particular problem in estuaries and deltas, where the runoff that was aggregated by watersheds is finally dumped at the mouths of major rivers.

Cause 3: Oil, Gasoline and Additives

Oil spills like the Exxon Valdez spill off the coast of Alaska or the more recent Prestige spill off the coast of Spain get lots of news coverage, and indeed they do cause major water pollution and problems for local wildlife, fishermen, and coastal businesses. But the problem of oil polluting water goes far beyond catastrophic oil spills. Land-based petroleum pollution is carried into waterways by rainwater runoff. This includes drips of oil, fuel, and fluid from cars and trucks; dribbles of gasoline spilled onto the ground at the filling station; and drips from industrial machinery.

Shipping is one of these non-spill sources of oil pollution in water: Discharge of oily wastes and oil-contaminated ballast water and wash water are all significant sources of marine pollution, and drips from ship and boat motors add their share. Drilling and extraction operations for oil and gas can also contaminate coastal waters and groundwater.

Cause 4: Mining

Mining causes water pollution in a number of ways:

- ☆ The mining process exposes heavy metals and sulphur compounds that were previously locked away in the earth. Rainwater leaches these compounds out of the exposed earth, resulting in "acid mine drainage" and heavy metal pollution that can continue long after the mining operations have ceased.

- Similarly, the action of rainwater on piles of mining waste (tailings) transfers pollution to freshwater supplies.
- In the case of gold mining, cyanide is intentionally poured on piles of mined rock (a leach heap) to chemically extract the gold from the ore. Some of the cyanide ultimately finds its way into nearby water.
- Huge pools of mining waste "slurry" are often stored behind containment dams. If a dam leaks or bursts, water pollution is guaranteed.
- Perhaps the worst offense in the category of mining vs. water pollution causes: Mining companies in developing countries sometimes dump mining waste directly into rivers or other bodies of water as a method of disposal. Developed countries are not immune from such insanity: The US government in 2003 reclassified mining waste from mountaintop removal (a type of coal mining) so it could be dumped directly into valleys, burying streams altogether.

Cause 5: Sediments

When forests are "clear cut," the root systems that previously held soil in place die and sediment is free to run off into nearby streams, rivers, and lakes. Thus, not only does clearcutting have serious effects on plant and animal biodiversity in the forest, the increased amount of sediment running off the land into nearby bodies of water seriously affects fish and other aquatic life. Poor farming practices that leave soil exposed to the elements also contribute to sediment pollution in water.

Cause 6: Chemical and Industrial Processes

Almost all bodies of water in the world have some level of pollution from chemicals and industrial waste.

Cause 7: Plastic

Plastics and other plastic-like substances (such as nylon from fishing nets and lines) can entangle fish, sea turtles, and marine mammals, causing pain, injury, and even death. Plastic that has broken down into micro-particles is now being ingested by tiny marine organisms and is moving up the marine food chain.

Sea creatures that are killed by plastic readily decompose. The plastic does not–it remains in the ecosystem to kill again and again.

Cause 8: Personal Care Products, Household Care Products and Pharmaceuticals

Whenever we use personal-care products and household cleaning products–whether they be laundry detergent, bleach, or fabric softener; window cleaner, dusting spray, or stain remover; hair dye, shampoo, conditioner, or Rogaine; cologne or perfume; toothpaste or mouthwash; antibacterial soap or hand lotion–we should realize that almost all of it goes down the drain when we do laundry, wash our hands, brush our teeth, bathe, or do any of the other myriad things that incidentally use household water. Similarly, when we take medications, we eventually excrete the drugs in altered or unaltered form, sending the compounds into the waterways. Studies have shown that up to 90 per cent of your original prescription passes out of you unaltered. Animal farming operations that use growth hormones and antibiotics also send large quantities of these chemicals into our waters.

Unfortunately, most wastewater treatment facilities are not equipped to filter out personal care products, household products, and pharmaceuticals, and a large portion of the chemicals passes right into the local waterway that accepts the treatment plant's supposedly clean effluent.

Study of the effects of these chemicals getting into the water is just beginning, but examples of problems are now popping up regularly:

- ✰ Scientists are finding fragrance molecules inside fish tissues.
- ✰ Ingredients from birth control pills are thought to be causing gender-bending hormonal effects in frogs and fish.
- ✰ The chemical nonylphenol, a remnant of detergent, is known to disrupt fish reproduction and growth.

Cause 9: Sewage

In developing countries, an estimated 90 per cent of wastewater is discharged directly into rivers and streams without treatment. Even in modern countries, untreated sewage, poorly treated sewage, or overflow from under-capacity sewage treatment facilities can send disease-bearing water into rivers and oceans. Leaking septic tanks and other sources of sewage can cause groundwater and stream contamination.

Cause 10: Air Pollution

Surprisingly enough, air pollution contributes substantially to water pollution. Pollutants like mercury, sulphur dioxide, nitric oxides, and ammonia deposit out of the air and then cause problems like mercury contamination in fish, acidification of lakes, and eutrophication (nutrient pollution). Most of the air pollution that affects water comes from coal-fired power plants and the tailpipes of our vehicles, though some also comes from industrial emissions.

Cause 11: Carbon Dioxide

According to the United Nations Environment Programme, a 15-year-long study of the role of man-made CO_2 in the earth's oceans found that the oceans had absorbed enough CO_2 to already have caused a slight increase in ocean acidification. The fear is that further CO_2 uptake will increase acidification even more and cause the carbonate structures of corals, algae, and marine plankton to dissolve. This could have significant impacts on the biological systems of our oceans.

Cause 12: Heat

Heat is a water pollutant–increased water temperatures result in the deaths of many aquatic organisms. These increases in temperature are most often caused by discharges of cooling water by factories and power plants.

Global warming is also imparting additional heat to the oceans. The impact on marine life is unknown at this point, but it's likely to be significant.

Cause 13: Noise

Many marine organisms, including marine mammals, sea turtles and fish, use sound to communicate, navigate, and hunt. The ever-increasing din of noise from ship engines and sonars has a negative effect. Because of this noise pollution, some species may have a harder time hunting; others may have a harder time detecting predators; still others may just not be able to navigate properly.

In a well publicized case in 2000, at least 17 whales were stranded on beaches in the northern Bahama Islands, with the likely cause being US Navy vessels operating mid-frequency sonar systems nearby.

Effects of Water Pollution

Effect 1: Water Borne Diseases

Human infectious diseases are among the most serious effects of water pollution, especially in developing countries, where sanitation may be inadequate or non-existent. Waterborne diseases occur when parasites or other disease-causing microorganisms are transmitted via contaminated water, particularly water contaminated by pathogens originating from excreta. These include typhoid, intestinal parasites, and most of the enteric and diarrheal diseases caused by bacteria, parasites, and viruses. Among the most serious parasitic diseases are amoebiasis, giardiasis, ascariasis, and hookworm.

Effect 2: Nutrient Pollution

The Woods Hole Oceanographic Institution calls nutrient pollution the most widespread, chronic environmental problem in the coastal ocean. The discharges of nitrogen, phosphorus, and other nutrients come from agriculture, waste disposal, coastal development, and fossil fuel use. Once nutrient pollution reaches the coastal zone, it stimulates harmful overgrowths of algae, which can have direct toxic effects and ultimately result in low-oxygen conditions.

Certain types of algae are toxic. Overgrowths of these algae result in "harmful algal blooms," which are more colloquially referred to as "red tides" or "brown tides." Zooplankton eat the toxic algae and begin passing the toxins up the food chain, affecting edibles like clams, and ultimately working their way up to seabirds, marine mammals, and humans. The result can be illness and sometimes death.

Nutrient-pollution-driven blooms of non-toxic algae and seaweed can also cause problems by reducing water clarity, making it hard for marine animals to find food and blocking the sunlight needed by sea grasses, which serve as nurseries for many important fish species.

When the algal overgrowths finally die, they sink to the bottom and begin decomposing. This process uses oxygen from the surrounding water. In some cases, the decomposition process takes enough oxygen out of the water that the level falls too low to support normal aquatic life and the region becomes a coastal dead zone.

Finally, nutrient pollution can trigger unusual outbreaks of fish diseases. For instance, scientists have found that Pfiesteria, a tiny marine pathogen, can thrive in nutrient-polluted waters.

Effect 3: Chemical Contamination

Over the years, many types of chemicals have gotten into our waterways–and they continue to do so today. Chemical water pollution typically occurs because:

1. The chemicals are dumped into the water intentionally;
2. The chemicals seep into groundwater, streams, or rivers because of failing pipes or storage tanks;
3. The chemicals catastrophically contaminate waterways because of industrial accidents;
4. The pollutants are settled out of polluted air (or are precipitated out of polluted air); or
5. Chemicals are leached out of contaminated soil.

The above types of chemical contamination are considered "point sources" of water pollution. Non-point-source chemical pollution also occurs via pesticide runoff from farm fields and homeowners' lawns, as well as runoff of automotive fluids and other chemicals from roads, parking lots, driveways, and other surfaces.

- ☆ Severe chemical spills and leaks into surface waters usually have an immediate effect on aquatic life (fish kills, etc.).
- ☆ The human effects of chemical pollution in water can generally be viewed the same as any other form of chemical contamination–water is just the delivery mechanism.

There are a few broad categories of water pollution effects related to chemicals that are worth exploring further, which we do below.

Pesticides

- ☆ Pesticides are carried in rainwater runoff from farm fields, suburban lawns, or roadside embankments into the nearest creeks and streams. Occasionally they are even intentionally sprayed into waterways as part of a pest-control effort.

- Pesticides can accumulate via water into the food chain as well, ultimately being consumed by humans or animals in food.
- In the most infamous case of pesticide pollution, widespread use of the insecticide DDT polluted waterways, contaminating fish, and ultimately poisoning bald eagles (and other animals) that ate the fish. DDE, the principal breakdown product of DDT, built up in the fatty tissues of female eagles and prevented sufficient calcium being released to produce strong egg shells. The thin shells would break when the parents sat on the eggs to keep them warm. DDT affected many other species as well.
- In terms of general human health effects, pesticides can.
  - Affect and damage the nervous system;
  - Cause liver damage;
  - Damage DNA and cause a variety of cancers;
  - Cause reproductive and endocrine damage;
  - Cause other acutely toxic or chronic effects.

Oil and Petroleum Chemicals

When oil pollution gets in water, some of the components of are degraded and dispersed by evaporation, photochemical reactions, or bacterial degradation, while others are more resistant and may persist for many years, especially in shallow waters with muddy sediments.

Though much scientific work remains to be done on the effect that petroleum pollution has on plants and animals, we do know a few things:

- Exposure to oil or its constituent chemicals can alter the ecology of aquatic habitats and the physiology of marine organisms.
- Scientists know that oil (or chemical components of oil) can seep into marsh and sub-tidal sediments and lurk there for decades, negatively affecting marsh grasses, marine worms, and other aquatic life forms that live in, on, or near the sediment.

- Evidence strongly suggests that components of crude oil, called polycyclic aromatic hydrocarbons (PAHs), persist in the marine environment for years and are toxic to marine life at concentrations in the low parts-per-billion range. Chronic exposure to PAHs can affect development of marine organisms, increase susceptibility to disease, and jeopardize normal reproductive cycles in many marine species.

Mercury

Mercury finds its way into water primarily through air pollution from coal-fired power plants and some other industrial processes. In the water, the elemental mercury is converted to methylmercury by certain bacteria, after which it moves up the food chain of fish gobbling each other up. In the end, the larger fish may end up on your dinner plate–swordfish, sea bass, marlin, halibut, or tuna, for example.

The effects of mercury on humans are already pretty well understood. Young children and foetuses are most at risk because their systems are still developing. Exposure to mercury in the womb can cause neurological problems, including slower reflexes, learning deficits, delayed or incomplete mental development, autism, and brain damage. Mercury in adults is also a problem, causing:

- Central nervous system effects like Parkinson's disease, multiple sclerosis, and Alzheimer's disease.
- Heart disease.
- In severe cases, causing death or irreversibly damaging areas of the brain.

Animals in any part of the food chain affected by the bioaccumulation of mercury can also suffer the effects of mercury pollution. Possible effects include death, reduced reproduction, slower growth and development, and abnormal behaviour.

Effect 4: Pollution Due to Mining

There are a number of negative water-pollution effects from mining operations:

Acid Mine Drainage

When rain or surface water flows over exposed rock and soil, it can combine with naturally occurring sulphur to form sulphuric acid. The acidified rainwater eventually finds its way to streams and groundwater, polluting them and impacting local aquatic life. Some streams can become so acidic–more acidic than car-battery acid–the aquatic ecosystem is completely destroyed. The same leaching process that causes acid mine drainage can impart heavy-metal pollutants from the soil and rock as well.

Spills and Leaks

Whether it's a leak in the containment system of a cyanide leach heap or a breach in a coal-slurry impoundment dam, the result is the same–pollution of streams, rivers, and groundwater, killing aquatic life and poisoning drinking water.

Mountaintop Removal Mining

In this technique, the tops of coal-rich mountains are removed and the overburden is dumped into nearby valleys, burying stream habitats altogether, with the obvious catastrophic effect on whatever life forms lived in or around the stream.

Effect 4: Marine Debris

Marine debris is basically trash in the ocean. Trash fouls inland waterways too, for sure, but trash seems to be a particular problem in our seas. The Ocean Conservancy calls marine debris one of the world's most pervasive marine pollution problems.

The debris includes escaped inland trash and garbage thrown overboard by ships and boaters–plastic bottles and bags, six-pack rings, cigarette butts, Styrofoam, etc. Marine animals can swallow the trash items, which often look similar to prey they would normally eat, or the trash item may have barnacles or other detectables attached and is inadvertently ingested with the food. For instance, sea turtles will eat a plastic bag believing it to be a jellyfish. The bag can cause an intestinal blockage and sometimes death.

A new and potentially devastating effect of marine debris is emerging. After years of degradation at sea, plastic breaks up. The plastic has not biodegraded but rather has disintegrated into very small pieces. Marine animals near the bottom of the food chain are now ingesting these teeny-tiny little pieces of plastic pollution. How far up the food chain the stuff will go is unknown.

Discarded or lost fishing gear–line, rope, nets–and certain trash items can get wrapped around marine animals fins or flippers, causing drowning or amputation. Marine debris can also degrade coral reefs, sea grass beds, and other aquatic habitats.

Effect 6: Thermal Pollution

It's easy enough to see how discharging the heated-up water from a power plant into a river could cause problems for aquatic organisms used to having their water home stay at a fairly specific temperature. Indeed, industrial thermal pollution is a problem for our waterways–fish and other organisms adapted to a particular temperature range can be killed from thermal shock, and the extra heat may disrupt spawning or kill young fish.

Additionally, warmer water temperatures lower the dissolved oxygen content of the water. That's a double-whammy to aquatic organisms, since the warmer water also causes them to increase their respiration rates and consume oxygen faster. All this increases aquatic organisms' susceptibility to disease, parasites, and the effects of toxic chemicals.

Global warming is imparting extra heat to our oceans, which have absorbed about 20 times as much heat as the atmosphere over the past half-century. The ocean is a complex system, and scientists don't know yet what all of the effects of this type of "water pollution" will be, but here are some likely ones:

- ☆ Sea levels will rise (because of thermal expansion and melting ice), increasing coastal flooding and inundation.
- ☆ There will be more intense hurricanes as they gather additional strength from warmer surface waters.
- ☆ Temperature-sensitive species like corals will see tougher times. Over the last 2-3 decades, temperatures in tropical waters have increased nearly 1 degree Fahrenheit. That may not sound like much, but it's been enough to increase cases of coral bleaching. A study in Science estimated that if carbon dioxide releases continue at the current rate, by mid-century ocean conditions (increased water temperatures + increased ocean acidity) will make it impossible for most corals to survive.

Effect 7: Noise Pollution

"Noise pollution" from ship engines and sonar systems make it difficult for marine mammals like whales, dolphins, and porpoises to communicate, find food, and avoid hazards. Powerful sonar systems operating at certain frequencies have been implicated in whale beachings and may cause damage to marine mammals' sound-sensitive internal structures, causing internal bleeding and even death.

Frequent or chronic exposure to both high- and low-intensity sounds may cause stress on all higher forms of marine life, potentially affecting growth, reproduction, and ability to resist disease.

Effect 8: Cost to Consumers

Tap water quality is regulated, and nearly 100 per cent of community water systems in the US are meeting clean drinking water standards. But is that good enough? Why are so many people convinced it's worth buying bottled water?

When the Environmental Working Group tested tap water from a number of cities, it found 119 "normal" chemicals–those for which the EPA has set health-based limits–and another 141 completely unregulated chemicals. If tap water has that many chemicals in it but is still classified as meeting water quality standards, one might say that the standards are, um, lax.

If nothing else, it's fair to say that even "safe" tap water usually has a "chloriney" taste. Chlorine and its disinfection byproducts are known health threats, and none of us wants to be the victim of the next cryptosporidium-in-drinking-water problem or some similar nightmare.

So it's no wonder many of us go to the extra expense and trouble of buying bottled water or using water filters, even though there are no regulations that guarantee those approaches will provide water that is safer.

Our first water pollution solution is simple: Enforce existing laws. A politician pontificating about a great new anti-pollution law they've sponsored means little if they continue to allow existing laws to go unenforced. Beyond laws, there are some practical water pollution solutions that can be implemented by society and by you as an individual.

Solutions to Water Pollution

Solution 1: Reducing Nutrient and Pesticide Pollution

Solutions to water pollution caused by excess nutrients and chemical pesticides can be found in five broad categories:

Encourage Smart Agricultural Practices

Right-sizing applications of fertilizer and using techniques like biodynamic farming, no-till planting, settling ponds, and riparian buffer zones can help keep polluted runoff from entering streams.

Reduce Urban/Suburban Runoff of Lawn Fertilizers and Pesticides

If we put "normal" fertilizer, pesticides, and other chemicals on your lawn, landscaping, and gardens, we are part of the water pollution problem. While we may find these products helpful, much of their volume is washed by rain or blown by air to nearby streams, ponds, and rivers. They also tend to degrade your soil over time, making our future gardening efforts that much more difficult and reliant on chemicals.

Prevent Further Destruction Wetlands, and Reestablish them wherever Possible

Both inland and coastal wetlands act to buffer surges in runoff and to filter pollutants from runoff and flows. Yet it has been standard practice in the US (and many other countries) to allow development concerns to almost always trump the value of "nature's services." It's time to get serious about preserving wetlands.

Solution 3: Reducing Sewage Pollution

If the toilet in your house were spewing its contents onto your bathroom floor, you would make it a very high priority to get the situation corrected. As societies, we should place the same priority on upgrading out-of-date or under-capacity sewage treatment plants that sometimes spew their contents into our waterways. It's also important to ensure that homeowners with septic fields are installing and maintaining their systems in a way that does not contaminate nearby groundwater or surface water.

More specifically, we should:

Get going on fixing outdated municipal water treatment plants.

If we flush, we may be part of the problem. Society needs to ante up whatever is necessary to fix inadequate sewage treatment systems.

Conserve water which helps reduce loads on septic systems and treatment plants, reducing the likelihood that they will send waste into our waters.

Solution 3: Improving Storm Water Management and Watershed Monitoring

Water pollution control is most appropriately addressed at the watershed level. As the saying goes, everyone lives upstream of somewhere else. What happens in someone's back yard and neighbourhood impacts every other part of the watershed system at lower elevations–all the way to the ocean.

One thing that works well to help control backyard and neighbourhood pollution is to implement urban and suburban storm water management strategies, including:

- ✰ Preserving undeveloped land to help soak up rains; and
- ✰ Constructing wetlands, stream buffer zones, and settlement ponds to allow contaminated runoff to undergo natural biological remediation before it gets into the watershed.

Solution 4: Stopping Deforestation

A healthy forest acts like a sponge to soak up the rains when they come, holding the water and filtering it before it makes its way to nearby streams, lakes, and rivers. When all the trees are cut down–clearcutting is still logging companies' preferred method of operation–the forest ecosystem dies and can no longer perform this service. Rain water rushes directly into streams, flowing over exposed soil, picking up and carrying sediment pollution into nearby waterways.

You can support healthy forests by (a) supporting efforts to ban clearcutting; (b) supporting "roadless rules" that keep logging roads out of pristine national forests; (c) making smarter lumber and paper choices for yourself and implementing methods to save paper. It's also worth noting that paper manufacturing is a highly polluting affair, and using paper sourced with high levels of "post consumer content" helps reduce pollution from the production of virgin paper.

Solution 5: Opoosting Coastal Development

Natural shorelines (and the wetlands usually found there) serve many purposes, from fish nurseries to absorption of hurricane impact to filtration of the river water entering the estuary. But in the US alone, more than 20,000 acres of these sensitive areas disappear each year. When houses, hotels, and resorts go up or other development occurs, not only are the wetland and coastal eco-services lost, but the human activity imparts many types of pollution to these sensitive coastal areas.

Coastal development is a significant problem for the oceans, but all forms of suburban sprawl chew up wetlands, forests, meadows, and other natural areas that help soak up rains and filter water before it enters streams and rivers. Supporting smart growth, urban redevelopment, and open space preservation is an important solution to water pollution.

Solution 6: Reducing Pollution from Oil and Petroleum Liquids

While it's true that a large amount of oil naturally seeps into the ocean from underground geological sources, marine life in the areas where this occurs have had eons to adapt to the conditions. Human-caused petroleum pollution invariably happens in much more sensitive areas, often with disastrous consequences.

The first-level solution to this type of water pollution is to stop letting so much oil and oil byproducts get into the water in the first place. Yes, we must reduce the occurrences of oil spills; but more importantly, we must reduce the amount of petroleum pollution getting into waterways from non-spill sources, which contribute far more to the problem than spills.

Governments and corporations can respond to both types of petro-pollution by:

- ☆ Quickening the pace of moving the world's tanker fleet to all double-hull ships.
- ☆ Tightening regulations governing maintenance and inspections of commercial ships, motor boats, and recreational water craft, which can leak oil and fuel into the water. (And maybe it's a good idea if we completely prohibit motorized craft on our drinking-water reservoirs!)

- ✩ Requiring filtration ponds and natural buffer zones around roads and parking lots to help keep runoff contaminated with oil and gas drips from getting into waterways.
- ✩ Doing more citizen education on the subject of how to keep oil out of our environment.

Solution 7: Reducing Mercury Emmision

The solution to mercury pollution in our waters is to solve the mercury pollution problem coming from the land. The technology exists to do a much better job of controlling mercury pollution, and predictive models show that reducing mercury emissions to the air will reduce mercury pollution in water and the subsequent contamination of fish. For existing coal-fired power plants, better scrubber technology can be applied to get 90 per cent of the mercury out of the emissions. For new coal-fired power plants, we should be insisting that the only type of allowed construction is "coal gasification," which allows *all* of the mercury pollution to be filtered out.

Solution 8: Cleanig Up Mining Practices

Some streams are so polluted from acid mine drainage that workers who wade into streams to take water samples must wear protective boots and gloves, otherwise the polluted water could cause skin blisters and sores. There are worse cases of pollution from mining–some bodies of water near mining operations are completely devoid of life. In addition to increased acidity, the water can become contaminated with heavy metals like cadmium, which are leached into waterways by the action of rainwater on exposed rock and earth. In gold mining, arsenic leach heaps may leak arsenic into groundwater or surface water.

Coal, metals, and other products produced by mining are part of what built modern society. But techniques exist for doing these operations more cleanly, and we owe it to our waters and ourselves to insist on it.

As consumers and citizens, there are a number of things we can do:

- ✩ Insist that regulatory agencies force industry to clean up long-abandoned but still-polluting mines.

- Tell our elected officials to prevent the siting of new mines where they are likely to cause water pollution problems.
- Demand that the mining industry stop the highly destructive coal mining practice of mountaintop removal mining, which often buries streams altogether.

Solution 9: Cleaning Up Chemical Pollution

First, there are some things we should expect from our government and corporations:

- Reinvigorate progress on cleaning up superfund sites, and reestablish the "polluter pays" principle. Taxpayers should not foot the bill for decades of industry abuses.
- Clean up polluted brownfield sites–sites not quite bad enough to make the superfund list, but still pretty bad–and promote their redevelopment.
- Eliminate all remaining industrial waste-water discharges to streams, enforcing a "zero emissions" policy for the wastewater from our factories.
- Upgrade water treatment plants so they can filter out chemicals and pharmaceuticals. Most plants do not handle either.
- Continue the fight to stop emissions of acid-rain chemicals (sulphur and nitrogen) that not only damage forests but also acidify lakes.

There are also some water pollution solutions we as consumers and citizens can implement when it comes to chemicals:

- Buy organic food.
- Start buying "green" household cleaners and personal care products.
- How important good nutrition, sleep, and low stress levels are to keeping you healthy–and pharma-free.
- Stop all use of chemical pesticides around your house and yard.
- If you have to dispose of old paint, varnish, or other DIY chemicals, check with your local government's environment or public works office to find out the safest way to do so.

Solution 10: Fighting Global Warming

In terms of water pollution, there are two main threats from global warming:

Ocean Acidification

As atmospheric CO_2 levels have risen, ocean CO_2 levels have risen even more, thus increasing the acidity level of the ocean. At a minimum, this trend will negatively affect organisms with shells, which may dissolve or become malformed if the pH drops low enough.

Ocean Temperature

As the planet warms, so does the ocean. All organisms in nature have limits to the temperature range in which they can exist. Increasing the temperature of the oceans will have varying–but likely negative–effects on ocean creatures.

There are plenty more ways in which global warming will impact water; for instance, less mountain snow pack and smaller glaciers will result in lower river flows for many regions during summer, and melting ice sheets in Greenland and Antarctica will change salinity levels and ocean flows, and will raise ocean levels, inundating coastal properties and ecosystems.

References

1. Singh, A.K. (2004). Arsenic contamination in groundwater of North Eastern India. In: *National Seminar on Hydrology with Focal Theme on "Water Quality"*, Held at National Institute of Hydrology, Roorkee during Nov. 22–23.

2. Twarakavi, N.K.C. and Kaluarachchi (2006). Arsenic in the shallow groundwaters of conterminous United States: assessment, health risks, and costs for MCL compliance. J. J. *Journal of American Water Resources Association*, 42 (2): 275–294.

3-4. Mukherjee, A., Sengupta M.K. and Hossain, M.A. (2006). Arsenic contamination in groundwater: A global perspective with emphasis on the Asian scenario. *Journal of Health Population and Nutrition*, 24(2): 142–163.

5. Chatterjee, Amit, Das, Dipankar, Mandal, Badal K., Chowdhury, Tarit Roy, Samanta, Gautam and Chakraborti, Dipankar (1995).

Arsenic in groundwater in six districts of West Bengal, India: The biggest arsenic calamity in the world. Part I: Arsenic species in drinking water and urine of the affected people. *Analyst*, 120(3): 643–651.

6. Meliker, Jaymie R. (2007). Arsenic in drinking water and cerebrovascular disease, *diabetes mellitus*, and kidney disease in Michigan: A standardized mortality ratio analysis. *Environmental Health Magazine*, 2: 4.
7. APHA, AWWA (1998). *Standard Method for Examination of Water and Wastewater*, 20th edn. American Public Health Association, Washington, D.C.
8. Rajurkar, N.S., Nongbri, B. and Patwardhan, A.M. (2003). Physico-chemical and bacteriological investigation of River Umshyrpy at Shillong, Meghalaya. *Ind. J. Environ. Health*, 45: 83–92.
9. Singh, A.K. (2004). Arsenic contamination in groundwater of North Eastern India. In: *Hydrology with Focal Theme on Water Quality*, (Eds.) C.K. Jain, R.C. Trivedi and K.D. Sharma. Allied Publishers, New Delhi, pp. 255–262.
10. Bhattacharjee, S., Chakravarty, S., Maity, S., Dureja, V. and Gupta, K.K. (2005). Metal contents in the groundwater of Sahebgunj district, Jharkhand, India, with special reference to arsenic. *Chemosphere*, 58: 1203–17.
11. BGS/DPHE (2001). In: *Arsenic Contamination of Groundwater in Bangladesh*, (Eds.) D.G. Kinniburgh and P.L. Smedley. Final report, BGS technical report WC/00/19. Keyowrth, U.K.: Brittish Geological Survey, 2001.
12. Biswas, B.K., Dhar, R.K., Samanta, G., Mandal, B.K., Faruk, I. and Islam, K.S., *et al.* (1998). Detailed study report of Samta one of the arsenic affected village of Jessore district, Bangladesh. *I. Curr. Sci.*, 74(2): 134–145.
13. Chakraborti, D., Biswas, B.K., Basu, G.K., Chowdhury, U.K., RoyChowdhury, T. and Lodh, D., *et al.* (1999). Possible arsenic contamination free groundwater source in Bangladesh. *J. Surf. Sci. Technol.*, 15: 180–188.
14. Chakraborti, D., Basu, G.K., Biswas, B.K., Chowdhury, U.K., Rahman, M.M. and Paul, K. *et al.* (2001). Characterization of

arsenic bearing sediments in Gangetic delta of West Bengal, India. In: *Arsenic Exposure and Health Effects,* (Eds.) W.R. Chappell, C.O. Abernathy and R.L. Calderon. Amsterdam: Book: Elsevier Science, p. 27–52.

15. Chakraborti, D., Rahman, M.M., Paul, K., Chowdhury, U.K, Sengupta, M.K. andLoth, D. *et al.* (2002). Arsenic calamity in the Indian sub-continent what lesions have been learned? *Talanta* 58: 3–22.

Chapter 23

The Concept of Environment in Perspective of Applied Ethics

A.K. Verma, Ahmad Ali, Mohammad Ali and Arshad Ali

Introduction

In the age of development of science and technology "Environmental Ethics' is becoming most important topic in the light of global problem which is present due to environment pollution.

The word Environment is divided into two parts "Environ+ment' which means 'encircle' or 'all around'. In Hindi it is known as "Paryavaran' and its meaning is also refers same thing *i.e.* 'it is an external cover which bounded the whole biological system.

World dictionary also suggests that Environment is the sum of condition, agencies and influences which affect the development, growth, life and death of an organism, species to race.

Environment is also called 'Ecology' which comes from Greek language 'Oikos' and 'logos'. It means home and house and its study respectively. So whole Ecology means the study of home or habitat. In a such way Ecology is the study of Environment of any organism.

Collins Cobuild English Dictionary defines Ecology as "The pattern and balance of relationship among plants, animals and people and their environment.

The Components of Ecology

There are four components of Ecology determiners and affect this system.

They are as following:

1. Abiotic substances
2. Producer organism
3. Consumer Organism
4. Decomposer Organism

Fields of Biosphere

1. Lithosphere
2. Hydrosphere
3. Atmosphere

Contents of Environmental

1. Human Ecology
2. Social Ecology
3. Population Ecology
4. Rural Ecology
5. Urban Ecology
6. Industrial Ecology
7. Water Ecology
8. Sky Ecology

The Main Principles of Environmental Ethics

1. Determinism
2. Possibilsm
3. Neo Determinism
4. Volunarism
5. Ecological Approach
6. Normative Approach

The Relationship of Human Between Environmental Ethics and Human Development

The relationship between Human and Environment have been changed from several decades and due to this we also find Environmental Ethics as many aspects like:

1. Hunting and food gathering period
2. Domestication of animals and pastoral period
3. Plantation and agriculture period
4. Science, Technology and Industrial period.

The Historical Research of Development of Environmental Ethics

Indian Philosophy

The oldest source of education of human *i.e.* veda also declared the top position to Environmental Ethics and its value. The Vaidic Saint of India and all over the world who used to worship the nature also understand this fact they accepted the whole universe as one Brahma and all power of universe like fire, water, air, sky and Earth are considered as God. In veda it is also said that "Dya Ma abhilekhi Antariksham Ma Hinsi : Prithivyal Sambhava"-*Shu. Ya. Ve.– 43.*

The early man were totally vegetarian. But after the relationship with Anarya like marriage relationship then nonvage and other activities of violence came into their life and this karm kanda was highly criticized in the veda and purana by educated and sage people. In the period of Epic there was a devotion present for Environment by Arya.

In Darshan also Environment ethics is highly regarded and this culture is even presented in modern and contemporary Indian thought too. The famous Indian scientist Jagdish Chandra Basu gave scientific regardless to Environmental ethics by discovering the life of plant and its sentiment. Jain darshan also gave their priority to this field.

Western Philosophy

In western philosophy the Greek philosopher like Sukrat, Plato Aristotle, Denocritics, Heraclits, etc. also paid their regard to environmental ehics. In the century of 17th to 19th western thinkers also started taking interest in Environmental studies and its components in which Reumo, Bufen, August, Greesvek, Mobices Stephen Foebs, J. Warming, B.E. Shellford, Charles Alten, Charles Darwin etc. are famous thinkers. In the time of modern age the famous western thinkers like Descarts Spinoza, Leibinitz, Kant have shown their respect to nature. On the other hand in contemporary age huministics thinkers like comte, Bradley, Aienstein, Newton, Alexender are also supported this through their ideas.

The Global Effort for Protecting Environmental Ethics

After second world war and failure of league of nations the United nations organization named global centre was established on 24th October 1945 for the sake of global peace and security. In this human rights were proclaimed on 10 December 1948. In those rights it was sworn to protect human respects, liberty, equality, justice etc. In 1972 in Sweden at stock home 119 countries proclaimed for protection of Environment in which 16 main points were taken. By this one Earth declaration was arranged and by the basis of this UNEP was organised. Again in 1975 in Belgrade IEEEP was organised. Not only this but in 1982 Human Environment International conference took place in Nairobi in which 105 countries were participated.

India also organised a committee named National Committee for Environmental Co-ordination (NCEPC). Under this following points are declared:

1. Control over industrial pollution.
2. Protection of losteal plants and animals.
3. Availability of pure water.
4. Protection of natural resources like land, water, plants, atmosphere.

In other country of world many program are started for protection of Environment as U.N.O.'s guide lines at political and nonpolitical stage. Due to this in present time animals rights, protection of plants, forest protection, control over water, air and sound pollution, eco friendly transportation has started as well as people are awaking towards global warming, contraction of glaciers, deforestation etc. Indian constitution adopted the protection of Environment for every people as fundamental work.

But every coin has other face too in the same way unfortunately someone also present in this society who says development is very important according to this age and we cannot walk as development. But they can even understand this fact that environmental ethics is not a big stone in the way of development and systematic development not development at any cost or hook, or by crook.

But in present world all such effort seems being fail under such drastic condition and problems. One thing must be essential that human should develop a vision of protection of environment and avoid the selfishness and also should avoid the mentality to owning every thing unless human will always far away from the idealism of Environmental Ethics which is most important for the purpose of living of human in this beautiful world.

References

1. Mishra, N. *Nitishashtra*
2. Mappes, T.A. and Zembaty, J.S. *Social Ethics: Morality and Social Policy.*
3. Singer, Peter. *Practical Ethics.*
4. Sasso, James. *Global Ethics.*
5. Kumari, Sunanda. *Boudh avam Jain Nitishashra.*
6. Verma, V. P. *Nitishashra ke Mul Sidhanta.*
7. Chaurasia, M.P. *Anuperyukta Nitishashra.*
8. Mishra, Jha Avam. *Achar Shashtra ke mul Sidhanta.*
9. Singh, B.N. *Nitishashra.*
10. Verma, A.K. *Nitishastra Ki Ruprekha.*

Chapter 24

Right to Clean Water: A Constitutional and Legislative Approach

H.N. Tiwari and Poonam

ABSTRACT

In ancient time our ancestor gave an important place to environment. Worship of nature- sun, moon, earth, air, water, tree and river etc was not merely primitive man's response to the fear of the unknown but arose from the deep reverence shown to the forces of nature which sustained and preserved human life on earth. One of the descriptions of water is "jeevan" means life.

Indian Constitution does not specifically mention "Right to Water" as a fundamental right. Nevertheless Right to life as provided under Art. 21 of the constitution contain Right to Water as fundamental right. The Supreme Court in Subhash Kumar v. State of Bihar had held that the right under Art. 21 include the right to enjoyment of pollution free water. Art. 48 A as directive principles and Art. 51 A(g) as fundamental duty has been incorporated in the constitution in order to protect environment. A number of statues such as the Water (Prevention and Control of Pollution) Act, the Water (Prevention and Control of Pollution) Cess Act 1977, the environment (Protection) Act 1986 and Rules frame there under lay down the guidelines for maintenance of water quality as well as for conservation of water. Since water is used for multifarious purposes it is an important component which falls within the domain of sustainable development and in order to preserve and protect it the polluter pays, precautionary principle and public trust plays significant role. In addition to water Act and Environment Act, constitution and IPC also lays down principles for preservation of water, *e.g.* Sec277IPC.

In this Chapter an attempt is made to examine the legal aspect relating to water especially clean water which is most sacared commodity and a large number of Indian populations are not in a position to secure safe drinking water and this is a regrettable thing in a welfare society.

Introduction

An examination of human history shows that human civilization flourished mostly on the banks of the rivers. Even today we find that most of the cities and particularly the metropolitan cities are situated on the banks of various rivers. The reason for this is very obvious; it is due to the availability of water. This shows the importance of water to the mankind from times immemorial. Water helps in sustaining life therefore; it must be preserved and protected from all types of pollutants. It must be free from every form of contamination so that it must fit for human consumption. The fast growing human population and rapid industrialization are leaving their impact on environment but it is more apparent on water. It is not only getting polluted on the surface of the earth even undergroundwater is not suitable for consumption without being treated to make it suitable. That is why the proverb- JAL HI JEEVAN HAI.

The constitution of India does not especially mention "Right to Water" as a fundamental right. Nevertheless right to life as provided under Article 21 of the constitution has been explained to contain "Right to Water" as a fundamental right. The Supreme Court in Subhash Kumar v. State of Bihar (1991) had held that the right under Article 21 of the constitution includes the right to enjoyment of unpolluted water particularly potable water.

Position during Ancient Period

In ancient time protection of nature/environment and preservation of natural resources were very much religious and an accepted mode of worshipping God. People did not see much difference between nature and God. In Rigveda, Atharveda and Yajurveda there are various verses in praise of Lord Varun (god of water) and Lord Indra. In manusmiriti, water is regarded as a creator and source of life on earth. It has further been mentioned in household rules of Manusmiriti that:

नाप्सु मूत्रां पुरीषं वाष्ठीवनं वा समुत्सृजेत्।

अमेध्यमलिप्तमन्यद्वा लोहतं वा विषणि वा।

It means "let him not through urine or faeces into the water, nor saliva, nor clothes defiled by impure substances nor any other impurity, nor blood, nor poisonous thing.[1]

From ancient to modern time water is the first priority of mankind and this continues even today inspite of much advancement in the field of science and technology. As water is the primary requirement for substance of human life and without which right to life cannot attained. It is hence implied to be a human fundamental right.[2]

Right to Clean Water in an International Prospective

International efforts for the protection and preservation of the global environment started with convening of the Stockhlom Conference to the Rio Summit led to the recognition that all human being are entitled to be a healthy and productive in harmony with nature.[3]

Internationally the Right to Water is not directly recognized although there is gathering momentum to do so. The adoption in 2002 of General Comment on the right to water by United Nation Committee on Economic, Social and Cultural Right is understood as the defining moment in supporting a human right approach, articulated as the right of everyone to sufficient, safe acceptable physically accessible and affordable water for personal and domestic uses. Subsequently, the UNDP's 2006 Human Development Report recommended that the countries should make water a human right.[4]

Principle 2 of the Stockholm Declaration on Human Environment 1972 require that "the natural resources of the earth including water, land, flora and fauna and especially representative samples of natural ecosystem must be safeguarded for the benefit of present and future generations through careful planning or management as appropriate." Significance of purity and sufficiency of water were also explicitly emphasized in the proclamation on Nov. 10, 1980 when United Nations declared that International Drinking Water Supply and Sanitation Decade. India is also a signatory to this declaration.[5]

Right to Water in a Constitutional Prospective

Constitution of India was established with the view to make India a welfare state. The State was to look after its citizens and provide them with the entire essential element necessary for their growth and development. Inspite of this the constitution does not explicitly mention the right to water as fundamental Right. The right to water has been construed by the judiciary as a part of Art. 14, Art. 21 and art. 39(b).[6]

Art.39(b) of the constitution which deals with the Directive Principles of State Policy states that " the State shall in particular direct its policy towards securing that the ownership and control of the material resources of the community are so distributed as best to subserve the common good.[7]

Although the Directive Principles of State Policy is non justifiable but they are fundamental to the governance of the country and it is the duty of the state to apply directive principles while making[8] Thus Art. 39(b) states that material resources of the community are so distributed which serve best purpose and for the welfare of the state and people. According to Entry56, Union List in 7th schedule of the constitution, the Central Government has authority over "Regulation of interstate rivers and rivers valleys to the extent to which such regulation and development under the control of the union is declared by Parliament by law to be expedient in public interest. Entry 17 of the State List in 7th Schedule to the Constitution gives the State government the power to control water supplies irrigation and canals, drainage and embankments water storage and water powers.[9]

In India there is marked difference between ownership of water for different resources, the water in the river, stream, lake *i.e.* surface water is considered to be the property of the State while water beneath the surface *i.e.* groundwater is considered to be the property of the land owner. This has led to unrestricted extraction of groundwater which happened in almost the whole of the country.[10]

Art. 48 A of Indian Constitution which is a part of the Directive Principles of State Policy deals with the protection of environment " the state shall endeavour to protect and improve the environment and to safeguard the forest and wildlife of the country.[11]

Art. 48A expressly direct the State to protect and improve the environment water is one of the component of environment. It is not only the duty of the state to protect and improve the environment which includes water also. Indian constitution also imposes duty upon the citizens of this country to protect and improve the environment. Art. 58A (g) states that "it is the duty of every citizen of India to protect and improve the natural environment including forest, lakes, rivers and wild life and to have a compassion for living creatures.[12]

Indian Constitution contains specific provisions for the protection and improvement of the environment but does not expressly mentioned right to water as fundamental right. "Right to life" as provided under Art. 21[13] of the Constitution does contain Right to Water as fundamental right. In Subhash Kumar v. State of Bihar[14] the Supreme Court had held that "the Right to enjoyment of pollution free water and air which include right to live". Water is the basic need for the survival of human beings and is a part of the right to life as enshrined in Art. 21 of the constitution and right to healthy environment and sustainable development are fundamental human right implicit in the right to life.[15]

Water is used for multifarious purposes it is an important component of environment which falls within the domain of sustainable development and in order to preserve and protect it the polluter pays, precautionary principle and doctrine of public trust play an important role.

In Vellore Citizen's Welfare Forum v. Union of India[16] the Supreme Court had held that while industries are of vital importance for the country's progress and development as they are generate foreign avenues, but having regards to the pollution caused by these industries, the principle of sustainable development has to be adopted as a balancing concept between ecology and development. Sustainable development as defined by the Brundtland Report means that meet the need of the present without compromising the ability of the future generations to meet their own need. Some of the salient principles of "Sustainable Development are Inter-Generational

Equity, Use and Conservation of Nature Resources, Environmental Protection, the Precautionary Principle, Polluter Pays principle, Obligation to assist and cooperate, Eradication of Poverty and Financial Assistance to the developing countries.

The 'precautionary principle' when applied by the courts to Indian condition means: (i) that environmental measures taken by the state and the statutory authorities must anticipate, prevent and attack the causes of environmental degradation; (ii) that where there are threats of serious and irreversible damage, lack of scientific certainty should not be used as a reason for posting measures to prevent environmental degradation; and (iii) that the 'onus of proof' is on the actor or the developer/industrialist to show that his action is environmentally benign.[17]

In Vellore citizen's welfare forum case the court directed the closure of those industries unless they install pollution control devices. Here the court also applies Polluter Pays Principles and impose a pollution fine of Rs 10,000 on each industries. The polluter pays principle is a popular concept which means that if a person causes any harm to environment or causes pollution, he is liable to clean it up and pay for it. The cost of causing damage to the environment is borne by the polluter

Another important doctrine incorporated by the Supreme Court in MC Mehta v. Kamal Nath[18] is the Doctrine of Public Trust. In this case Supreme Court held that motel interfered with the natural flow of the river by trying to block the natural relief spill of the river and thus directed cancellation of lease deed. The court observed that the public trust doctrine primarily rest on the principles that certain resources like air, water, sea and forest etc. have such a great importance to the people as whole that it would be wholly unjust to make them a subject of private ownership.

Right to Clean water: A Legislative Approach

There are various statues enacted by the Parliament which lay down the guidelines for the maintenance of water quality as well as for conservation of water. According to the Report of Tiwari committee, there are over two hundred central and marriage legislation that have some bearing on environment protection in most cases the environment concern is incidental tp the law's principal object.[19]

The Water (Prevention and Control of Pollution) Act 1974

An Act to provide for the prevention and control of water pollution and the maintaining or restoring of wholesomeness of water, for the establishment, with a view to carrying out the purposes aforesaid, of Boards for the prevention and control of water pollution, for conferring on and assigning to such Boards powers and functions relating thereto and for matters connected therewith.[20]

Thus the main purpose of this Act is to provide and control the pollution of water. Alongwith this objective, it also tries to maintain or restore wholesomeness of water and constitute boards for the aforesaid purposes.[21]

The Act provides for a permit system or consent procedure to prevent and control the water pollution. This Act generally prohibits disposal of polluting matter in stream, well and sewers or on land in excess of the standard established by the state Board. A person must obtain consent from the state board. Before taking steps to established any industry, operation or process or any addition to such a system or any extension or any addition to such a system which might result in the discharge of sewage or trade effluent into a stream, well or sewer or onto land. Along with this there are various other functions which are perform by the state boards:

1. Planning a comprehensive programme for prevention control and abatement of water in the state,
2. Encouraging, conducting and participating in investigation and research of water pollution problem
3. Inspecting facilities for sewage and trade effluent treatment
4. Developing economical and reliable methods of treatment of sewage and trade effluents.[22]

The Central Board may advice the control government on water pollution issues; coordinate the activities of state pollution control board, sponsor investigation and research relating to water pollution.

The Water (Prevention and Control of Pollution) Cess Act 1977

This Act was passed to meet the expenses borne by the central water board and state water board. This Act provides for levy and collection of the cess on water consumed by person carrying on certain industries and local authorities[23].

The Environment (Protection) Act 1986

The main object for the enactment of this act is to provide for the protection and improvement of environment and for the matter connected therewith. Whereas decision were taken at United Nations conference on Human Environment held at Stockholm in June 1972, in which India participated, to take appropriate steps for the protection and improvement of Human Environment. And whereas it is considered necessary further to implement the decisions aforesaid in so far as they relates to the protection and improvement of environment and prevention of hazards to human beings other living creatures plants and property.[24]

This Act clearly extends to water quality and control of water pollution. Section 2(a) of the Act defines the environment to include water, air and land and interrelationship which exists among and between water and human being other living creatures, plants, microorganism and property. The Act authorizes the central government to establish standards for the quality of the environment and for emission or discharge of environment pollutions from any sources. The Ministry of Environment and Forest has published Environment (Protection) Rules establishing general standards and industry based standards for certain for certain types of effluent discharge. The ministry has not yet promulgated rules establishing ambient inland water quality standards through state boards must have regarded to the assimilative capacity of receiving bodies. This Act includes a citizen initiative provisions and a provision authorizing the central government to issue direct orders to protect the environment. The central government may delegates specified duties and power under the Environment Act to any officer, state government or other authority.[25]

Indian Easement Act 1882

An easement is the right which owner or occupier has for the beneficial enjoyment of land. The following has been held to easement-

1. Right to pollution free water
2. Right to flow of natural stream
3. Right to riparian owner to use water on his land.[26]

Under Sec. 7 of Indian Easement Act 1882 every riparian owner has the right to continued natural flow of the waters of natural its natural stream in its natural condition without obstruction or unreasonableness.[27]

Indian Penal Code

In order to protect from pollution, section 277 of IPC provides that Whoever voluntarily corrupts or fouls the water of any public spring or reservoir, so as to render it less fit for the purpose for which it is ordinarily used, shall be punished with imprisonment of either description for a term which may extend to three months, or with fine which may extend to five hundred rupees, or with both.[28]

Conclusion and Suggestions

Water, which is an essential component of nature or environment, is a precious gift of nature and very essential not only to the mankind but also for all living beings. But it is useful when it is unpolluted and uncontaminated. There are prophecies that there is a probability that the third world war may take place for controlling water. The right to clean drinking water is not expressly mentioned neither in the constitution nor in any other legislation. The judiciary has played an important role and interpreted right to unpolluted or uncontaminated water as a Fundamental Right under Art. 21 of the constitution. Right to clean and pollution free water should be made an independent fundamental right.

The industries and urbanization is not only responsible for the pollution of water and its scarcity, general public is also responsible for the same because they wastewater without any strong reasons. Public and private sector participation is necessary for the preservation and conservation of pollution free water. In various part of the country undergroundwater is polluted and exploited, there is not only depletion of the reserve stock of undergroundwater, but it is also becoming contaminated day by day. In India another problem is that all laws deal with the surface water problem, an adequate and comprehensive legislation should be enacted which deals with the surface as well as undergroundwater.

Suggestions

1. Right to clean pollution free water should be made an independent fundamental right.

2. New effective technology should be adopted for making environment clean in general and water in particular.
3. Public awareness through media is necessary for their right *i.e.* Right to Clean Drinking Water.
4. Public and private sectors participation is necessary for preservation and conservation of water.
5. A comprehensive Act should be enacted by Parliament for the protection and conservation of undergroundwater.
6. National Water Policy should give adequate place for conservation of surface as well as undergroundwater resources.

References

1. Shastri, S.C. *Environmental Law*, 3rd edn. Eastern Book Company.
2. Dash, Satya Prakash. *Water: A Human Rights Perspective*. http: / /www.ielrc.org/activities/workshop_ 0612/content/ d0604.pdf
3. Leelakrishnan, P. *Environmental Law in India*, 3rd edn. Lexix Nexis Butterworths Wadhawa.
4. Narain, Vrinda. *Water as a Fundamental right: A Perspective from India*. http: //www.vjel.org/docs/Narain_Water_Draft. pdf
5. Supra note 1
6. *Right to Water: A Legal Prospective*. http: //chitranet.org/PDF/ Right_to_Water_and_the_law.328175840.pdf
7. *Constitution of India*
8. Art. 37 *Constitution of India*.
9. Supra note vi
10. Supra note vi
11. Supra note vii
12. Supra note vii
13. *Constitution of India*, Art. 21, Protection of life and personal liberty– No person shall be deprived of life and personal liberty except according to the procedure established by law.
14. AIR 1991

15. *Narmada Bachao Andolan* v. Union of India AIR 2000
16. AIR 1996
17. Batra, Majula. *Water Rights,* http: / /www.india–seminar.com/ 2000/492/492 per cent 20m. per cent 20batra.htm
18. AIR 1997
19. Diwan, Shyam and Rosencranz, Armin. *Environment Law and Policy,* 2nd edn. Oxford University Press.
20. The Water (Prevention and Control of Pollution) Act, 1974
21. Tiwari, H.N. *Environmental Law,* 3rd edn. Allahabad Law Agency.
22. Supra note xix
23. The Water (Prevention and Control of Pollution) Cess Act 1977
24. The Environment Protection Act 1986
25. Supra note xix.
26. Chitale, Atul. *Water for all Constitutional and Legal Imperatives.* http: / /ahec.org.in/wfw/web per cent 20ua per cent 20water per cent 20for per cent 20 welfare water/Atul_Chitale.pdf
27. Supra note xix
28. Indian Penal Code.

Index

www.ingramcontent.com/pod-product-compliance
Lightning Source LLC
Chambersburg PA
CBHW070710190326
41458CB00004B/927
9789351242307